Testing Quantum Mechanics on New Ground

Recent technological advances have made it possible to perform experiments, once considered to be purely *gedanken*, which test the counterintuitive and bizarre consequences of quantum theory. This book provides simple accounts of these experiments and an understanding of what they aim to prove and why this is important.

After introducing the main theoretical concepts and problems with the foundations of quantum mechanics, early chapters discuss experiments in the areas of wave-particle duality, cavity quantum electrodynamics and quantum nondemolition measurement. The text then examines investigation of new predictions, including the Aharanov–Bohm effect, before tackling the problem of macroscopic quantum coherence. Later chapters consider methods of testing the quantum Zeno paradox, collapse, macroscopic quantum jumps, tunneling times and Einstein–Bell nonlocality. Introductions to the theory behind new types of measuring devices such as micromasers and those based on the concept of quantum nondemolition are also given. Detailed references are included.

Suitable for non-specialists, this book will appeal to graduate students and researchers interested in the physics and philosophy of quantum theory.

PARTHA GHOSE was born on February 5, 1939. He received a D. Phil (Science) from Calcutta University where he worked under the supervision of S. N. Bose. He has since taught at the University at Visva-Bharati, Santiniketan and Calcutta University and has been involved in the organisation of BCSPIN Summer Schools. He was Visiting Researcher at Syracuse University in October 1994 and Visiting Professor at the Tata Institute of Fundamental Research in May 1996. He is a Fellow of the West Bengal Academy of Science and Technology, and was the Chairman of the Indian Physics Association (Calcutta Chapter) and the Vice President of the Indian Physical Society for one term. He is an Honorary Member of the British Empire. Professor Ghose is best known for his work with Dipanker Home and G. S. Agarwal on the nature of wave–particle duality of single-photon states. His previous works include *Riddles in Your Tea Cup* published by IOP Publishing. He won the National Award from the National Council for Science and Technology Communication, Government of India, for the best media coverage in science and technology during the period 1986–90.

Testing Quantum Mechanics on New Ground

PARTHA GHOSE

S. N. Bose National Centre for Basic Sciences, Calcutta

CAMBRIDGE
UNIVERSITY PRESS

CAMBRIDGE UNIVERSITY PRESS
Cambridge, New York, Melbourne, Madrid, Cape Town, Singapore, São Paulo

Cambridge University Press
The Edinburgh Building, Cambridge CB2 2RU, UK

Published in the United States of America by Cambridge University Press, New York

www.cambridge.org
Information on this title: www.cambridge.org/9780521554633

First published 1999
This digitally printed first paperback version 2006

A catalogue record for this publication is available from the British Library

Library of Congress Cataloguing in Publication data

Ghose, P. (Partha)
Testing quantum mechanics on new ground / P. Ghose.
p. cm.
Includes bibliographical references and index.
ISBN 0 521 55463 2
1. Quantum theory. I. Title.
QC174.12.G49 1999
530.12–dc21 98–38620 CIP

ISBN-13 978-0-521-55463-3 hardback
ISBN-10 0-521-55463-2 hardback

ISBN-13 978-0-521-02659-8 paperback
ISBN-10 0-521-02659-8 paperback

Contents

Preface

Careful experiments with radiation, molecules, atoms and subatomic systems have convinced physicists over the years that the laws governing them (embodied in quantum mechanics) are quite different from those governing familiar objects of everyday experience (embodied in classical mechanics and electrodynamics). Quantum mechanics has turned out to be a very accurate and reliable theory though, even after more than seventy years of its birth, its interpretation continues to intrigue physicists and philosophers alike. Thanks to enormous technological advances over the last couple of decades, it has now become possible actually to perform some of the *gedanken* experiments that the pioneers had thought of to highlight the counterintuitive and bizarre consequences of quantum theory. So far, quantum mechanics has emerged unscathed in every case, and continues to defy all attempts at falsification. The spectacular success of its working rules has spurred physicists in recent times to grapple seriously with its foundational problems, leading to new theoretical and technological advances.

On the other hand, general relativity, the paradigm of classical field theory, continues to remain as accurate and reliable as quantum mechanics in its own domain of validity, namely, the large-scale universe. To wit, the agreement between the predictions of general relativity and observation of the energy loss due to gravitational waves emitted by binary pulsars is just as impressive as the agreement between the prediction of quantum electrodynamics and the measured value of the Lamb shift in atoms. However, all attempts at quantizing general relativity have failed so far, leading some theorists to believe that the two theories are essentially incompatible. If this is true, clearly both must yield somewhere if we are ever to find a more general theory which encompasses them. The question is: where? This is why testing quantum mechanics on new ground has acquired added significance.

The real 'mystery' of quantum mechanics lies in the quantum theory of measurement. All problems of interpretation are closely related to it. The quantum theory of measurement splits the empirical world into two parts: (1) the system S to be observed and (2) the apparatus A (which includes the state preparation and registration devices). Some interpretations of quantum mechanics also include the environment E (the rest of the universe) and even the consciousness of the observer in the description of the measuring process. According to Bohr (1949),

> The main point here is the distinction between the *objects* under investigation and the *measuring instruments* which serve to define, in classical terms, the conditions under which the phenomena occur.

The splitting of the world into S and A, inevitable for an objective description of the physical system, and the necessity of using classical language for A resulted in Bohr's 'complementarity interpretation', according to which,

> However far the phenomena transcend the scope of classical physical explanation, the account of all evidence must be expressed in classical terms. ...Consequently, evidence obtained under different experimental conditions cannot be comprehended within a single picture, but must be regarded as *complementary* in the sense that only the totality of the phenomena exhausts the possible information about the objects.

These were the early attempts at defining and solving the interpretational problems of quantum mechanics that eventually led to their rigorous axiomatic formulation by von Neumann and Pauli in 1933. This interpretation accepts the universal validity of quantum mechanics and treats the measuring apparatus also as a quantum system. Every interaction of the system S and the apparatus A makes them evolve into a superposition $\Sigma_i c_i A_i S_i$, where A_i and S_i are basis states. The possible states of the system A_i are therefore correlated with the 'pointer states' A_i of the apparatus. A measurement consists in only one of these possibilities $A_n S_n$ being realized with probability $|c_n|^2$. Such a process cannot be realized by pure quantum mechanical evolution but only through 'state vector reduction' or 'collapse' which is non-unitary and irreversible. The deep conceptual problems inherent in this interpretation were dramatized by Schrödinger through his famous cat paradox in which the quantum indeterminacy of a radioactive substance is passed on to a macroscopic cat which becomes a linear superposition of a 'live' and a 'dead' cat, which is then *resolved* by direct observation'.

Another deeply disturbing aspect of quantum mechanics is the EPR paradox. In 1935 Einstein, Podolsky and Rosen produced a famous argument to show that the interpretation of quantum mechanics advocated by Bohr and his Copenhagen school was incomplete if certain reasonable assumptions were made concerning 'reality' and 'locality' against which there was not a shred of empirical evidence in those days. Bohr issued a denial and was declared the winner. The debate persisted at a philosophical level until 1964 when Bell produced his famous inequality based on *local realism* (i.e., locality plus reality as defined by Einstein, Podolsky and Rosen) which quantum mechanics violates. At last the issue was brought down from its philosophical height to the empirical level. But one had to wait until 1982 for a real experimental verdict. The ingenious experiments carried out by Aspect and his colleagues with correlated photons once again seemed to vindicate quantum mechanics.

After the appearance of the EPR argument and Bohr's reply, the Copenhagen school had to change its stance. They had to abandon the idea that every measurement caused an unavoidable 'disturbance' of the system measured. In fact, Bohr admitted that in a case like the correlated EPR pair, there was 'no question of a mechanical disturbance of the system under investigation'. Yet, he reiterated, 'there is essentially the question of *an influence of the very conditions which define the possible types of predictions regarding the future behaviour of the system*'. Interest in interpretational issues henceforth shifted to exploring the possibilities of establishing single-system interpretations of quantum mechanics considered as a universally valid theory. A plethora of different interpretations followed, the main concern being to explain within the general framework of the Copenhagen philosophy how the classical nature of a measuring apparatus with definite 'pointer positions' emerges from a quantum mechanical substratum. Everett's 'many worlds interpretation', the 'decoherence' approach of Zurek, Joos and Zeh, and the Gell-Mann–Hartle 'many histories' interpretation are all attempts along this direction. On the other hand, Bohm's causal and ontological interpretation attempts to solve the measurement problem by supplementing the wave function by nonlocal hidden variables that remove the quantum indeterminacy of the position variable. Others like Ghirardi, Rimini, and Weber modify the Schrödinger evolution by adding a new term that causes rapid spontaneous localization of objects that are macroscopic in the sense of containing a sufficiently large number of particles. Thus, the measuring device continues to remain quantum mechanical but spontaneously collapses to definite 'pointer states' because of its macroscopic nature.

The purpose of this monograph is not to give a comprehensive and state-of-the-art review of the theoretical and philosophical arguments concerning interpretations and the quantum measurement problem. Rather, the intention is to provide for non-specialists simple accounts of the new experiments and an understanding of why they are interesting and important. They would presumably like to know which aspect of quantum mechanics each experiment is testing, and why a theorist might possibly doubt this aspect. Are there alternative theories which make a different prediction? Is it possible to devise experiments that will distinguish between different interpretations of quantum mechanics? How can one quantify the extent to which an experiment confirms quantum mechanics? These are the sorts of questions that will be raised and answered whenever possible.

There is a common theme running through the book – experimental tests of quantum mechanics. Among the topics included are the tests of new predictions of quantum mechanics such as the Aharonov–Bohm effect and its progenies such as the Anandan–Aharonov–Casher effect as well as the theory behind new types of measuring devices such as micromasers and those based on quantum nondemolition measurements (to construct gravitational antennas, for example). Also included are accounts of the experiments that have been carried out to test the conceptual and foundational aspects of quantum mechanics, such as self-interference, the nature of wave–particle duality, coherence of macroscopic objects, spontaneous collapse, the quantum Zeno effect, macroscopic quantum jumps through interrupted fluorescence, tunneling times and, of course, Einstein–Bell nonlocality.

Certain topics that are of great current interest, such as quantum computation, quantum cryptography and teleportation have been omitted from this book only because they are still in a nascent and rapid stage of development, and any account of them is bound to get dated fairly quickly. Another topic that has also been omitted for a similar reason is Bose–Einstein condensation which was predicted in 1924/25 but was only recently directly observed. Its implications are wide ranging and are still being worked out.

The book was planned in consultation with Dr Dipankar Home to whom I am most grateful not only for his valuable suggestions but also for providing many references. I am also grateful to Professor C. K. Majumdar, Director, S. N. Bose National Centre for Basic Sciences, for granting me sabbatical and study leave to write the book. To Professor V. P. Gautam, Head, Theoretical Physics Department, Indian Association for the Cultivation of Science, I owe a special debt of gratitude for providing me library and other facilities without which it would not have been possible for me to undertake this task. I also spent a couple of weeks at the Tata Institute of Fundamental Research, Mumbai, collecting materials for the book, for which I am grateful to Professor V. Singh, Director, and members of the Theory Group. Thanks are also due to the authors who have permitted me to use figures from their published papers and to the editors of the journals in which they appeared.

Partha Ghose
Calcutta

Copyright acknowledgements

I am grateful to the authors of all the papers/articles in books from which a number of figures have been used for giving me the necessary permission. The specific sources have been acknowledged in the relevant figure captions. I would also like to thank the following publishers for granting me the required copyright clearances:

American Physical Society for figures 1.5, 1.6, 1.11, 1.12, 2.6 to 2.12, 4.2 through to 4.10, 6.1, 6.2, 8.1 to 8.7, 9.1 to 9.3, 10.3 and 10.4.

Nature (Macmillan Magazines Limited) for figures 1.2 (1991), 10.1 and 10.2 (1989); and

Physics Letters A for figures 1.7, 1.9, 1.10, 1.18 to 1.23 and 7.1.

Partha Ghose
Calcutta

1
Wave–particle duality

1.1 Introduction

Wave–particle duality was born with Einstein's famous 'light-quantum hypothesis' in 1905. The earlier introduction of Planck's constant in 1900 was done with a certain discretion regarding the necessity to change the basic laws of microscopic phenomena themselves. Planck was certainly not in favour of introducing any discreteness into the structure of the electromagnetic field. This somewhat diminished the significance of his own discovery. It was left to Einstein to discover that for the mean square of the energy fluctuation ϵ of the part of radiation lying within the frequency range v and $v + dv$ in a small partial volume V of a hole filled with radiation in thermodynamic equilibrium, Planck's radiation formula gives

$$\bar{\epsilon^2} = hvE + \frac{c^3}{8\pi v^2 \, dv} \frac{E^2}{V} \qquad (1.1)$$

if E is the mean energy of the radiation in V. The second term can be easily interpreted with the help of classical wave theory as due to the interference between the partial waves. However, the first term is in contradiction with classical wave theory. Einstein suggested that it could be interpreted, by analogy to the fluctuations of molecules in an ideal gas, as arising from the fluctuations in the number of 'light-quanta' of energy $\hbar v$ which remain concentrated in limited regions of space and behave like *independent* particles. He also showed that if one starts from the simplified Wien limit of Planck's law which holds for $hv \gg kT$, one obtains only the first term. It was in this way that Einstein was led to believe in his 'light-quantum hypothesis' in 1909. He had already guessed it from entropy considerations of Planckian radiation in 1905 and applied it at once to explain the photo-electric effect and Stokes' law of fluorescence, and later the generation of secondary cathode rays and to predict the high-frequency

1

limit of *Bremsstrahlung* [1].* This was the beginning of the notion of the
wave–particle duality of light. According to Einstein's ideas, 'a ray of light
expands starting from a point, the energy does not distribute on ever
increasing volumes, but remains constituted of a finite number of energy
quanta localized in space and moving without subdividing, and unable
to be absorbed or emitted partially'. If the localized quanta carry all the
energy and momentum, what happens to the electromagnetic waves which
can produce interference? The problem was so acute that Einstein referred
to these waves as *Gespensterfelder* (ghost waves) guiding the photons.
However, it was soon clear that such statistically independent light-quanta
could not be consistent with Planck's radiation law, which made people
skeptical of them [2]. The next step was taken in 1923 by Louis de Broglie.
In his doctoral thesis he proposed, by analogy to the wave–particle duality
of radiation proposed by Einstein and on the basis of the special theory of
relativity, that material particles ought also to posses a wave-like character.
de Broglie's hypothesis was based on his philosophy of physical realism
[3]. His basic idea was that each massive particle exists permanently as an
energy point within a surrounding wave packet. The wave precedes and
guides ('pilots') it along real trajectories. To be consistent with the special
theory of relativity, he also assumed that the particle vibrates and evolves
in such a way that it remains in phase with its 'pilot wave'. This guarantees
that every inertial observer will see the relativistic duo move together with
the same velocity. Einstein came to know of this thesis from de Broglie's
supervisor, Paul Langevin, but reserved his judgement until sometime in
the latter half of 1924. By this time he had received S. N. Bose's paper
which contained a novel derivation of Planck's law of radiation in which
Bose had treated radiation in thermal equilibrium as a gas of light-quanta,
but the statistical method he employed to count equally probable states
of such quanta was the one that Planck had used (most probably in
December 1900) for his material oscillators 'in an act of desperation' in
order to give a theoretical derivation of the law he had discovered earlier
purely empirically. This statistical method however implied that the light-
quanta could not be independent as Einstein had been led to believe. At
last the contradiction between light-quanta and Planck's law was resolved
– light-quanta were not like classical particles but were bosons, to use more
modern terminology. Einstein was delighted at this, and immediately set
forth to apply Bose's counting method to the quantum states of an ideal
gas of material particles. This was a daring step, and it paid rich dividends.
Einstein calculated the density fluctuations in a small volume of such a gas
and again found two terms, one corresponding to density fluctuations that

---

* Numbered reference list begins on p. 186.

one would expect from classical particle theory and one that contradicted it. There is evidence that Bose had also independently done the same calculation and found the same result [2] although he did not publish it. It was clear that this latter term could only be explained as the effect of interference of waves. Einstein immediately realized that de Broglie's idea of matter waves was not a mere analogy but had a deeper foundation. He suggested that it should be immediately investigated by looking for diffraction of molecular beams. Einstein wrote three papers in late 1924 and early 1925 elaborating these ideas. Schrödinger who was interested at that time in the quantum theory of ideal gases studied them and eventually came up with the equation that such matter waves must satisfy. That was how wave mechanics was born in 1926. The empirical verification of the wave–particle duality of matter came shortly after from the experimental discoveries of Davisson and Germer, and G. P. Thomson [2]. However, as we will see, the real verification of the wave–particle duality of light, the progenitor, had to wait until the latter part of the 1980s when the development of more advanced technology made it possible to produce 'single-photon states' and their detectors.

1.2 The double-slit experiment

The double-slit experiment with single particles passing through the apparatus one at a time, lies at the heart of all debates regarding interpretations of quantum mechanics and the measurement problem. Historically it figured first at the famous Einstein–Bohr discussions during the 1927 Solvay Conference in Brussels [4]. Initially Einstein attempted to prove the incompleteness of quantum mechanics by trying to demonstrate that one could construct the interferometer in such a fashion that one could determine the entire path taken by each particle (which hole it passes through) and at the same time observe the self-interference pattern build up on the screen by the successive arrivals of the particles over a sufficiently long exposure time. The device he had in mind is shown in Fig. 1.1. The screen with the aperture (a horizontal slit of appropriately narrow width) is suspended by means of weak springs from a solid yoke bolted to the support on which other immobile parts of the apparatus are also attached. Every time a particle passes through the aperture, it imparts a momentum to the screen that can be accurately measured and the path taken by the particle determined by applying the laws of conservation of energy and momentum to the process. In this way it is possible to determine which of the two open holes in a second screen (not shown) a particle has passed through. Yet, quantum mechanics asserts that with both holes in the second screen open it is impossible *in principle* to de-

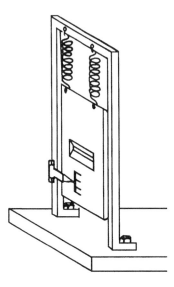

Fig. 1.1.   Einstein's thought experiment with a weakly suspended slit (after Ref. [4]).

termine which path a particle takes. Thus a more complete description of the events is possible than quantum mechanics can provide. Bohr refuted this argument by applying the position–momentum uncertainty relation to the first screen with the horizontal aperture, that is, by treating the screen quantum mechanically rather than classically as required of a measuring device! Herein lies the heart of the quantum measurement problem – *there is an inherent ambiguity in drawing a sharp dividing line between the quantum system and the classical measuring apparatus.* We will discuss this problem in greater detail in Chapter 5.

The separation Δx of the first maximum of the interference pattern from the central maximum on the third screen is related to the difference between the momenta of the particles reaching the screen through the two slits by $\Delta k \approx 2\pi/\Delta x$ (Fig. 1.2). Therefore, to be able to tell through which slit the particle reached the screen, the recoil momentum of the first plate must be determined with an accuracy $\Delta p \leq \hbar\Delta k \sim \hbar/\delta x$ where δx is the uncertainty in the position of this plate. This implies that $\Delta k > 1/\delta x$ and hence $\delta x \geq \Delta x$, which states that the uncertainty in locating the slits (and therefore the fringes) is larger than the spacing between the fringes, and the fringes are washed out. Therefore, there will always be 'a reciprocal relationship between our knowledge of the position of the slit and the accuracy of the momentum control' of the plate which will wash out the fringe pattern. The only way to determine which path a particle takes is to close one of the two holes in the second screen, in

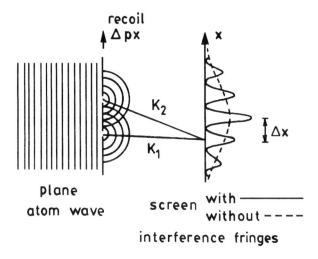

recoil
Δp_x

K_2

K_1

x

Δx

plane
atom wave

screen with ———
without – – – –

interference fringes

Fig. 1.2. Analysis of Einstein's thought experiment, using the uncertainty relation (after Ref. [5]).

which case the interference pattern necessarily disappears. In other words, interference (wave-like information) and a knowledge of 'which path' a particle takes all the way from the aperture to the final screen (particle-like information) are mutually exclusive. This is the essential content of Bohr's 'complementarity principle' [4] as applied to wave–particle duality. Phrased differently, it states that a description of the double-slit experiment in terms of classical pictures (waves *or* particles) is possible provided the condition of the apparatus is specified (both the holes open or one of them shut), and then a description in terms of *either* waves (when both holes are open) *or* particles (when only one hole is open) is possible.

Although the simultaneous observation of wave and particle behaviour is thus usually believed to be prohibited (the complementarity principle) by the position–momentum uncertainty relation, one can show that it is possible in practice to obtain 'which path' or particle-like informa-tion without scattering or otherwise introducing large uncontrolled phase factors into the interfering beams by using recent advances in quantum optics, namely the micromaser and laser cooling [5]. Einstein's goal is therefore indeed achievable. However, the interference fringes disappear once this path information is obtained. *This is because complementarity in interference experiments is guaranteed by the formalism of quantum me-chanics.* The wave function describing the particle (or any system) with an internal degree of freedom like, say, spin is given in the interference region by the sum of two terms referring to the two slits,

$$\psi(r) = \frac{1}{\sqrt{2}} [\psi_1(r)\chi + \psi_2(r)\chi] \equiv \frac{1}{\sqrt{2}} [\phi_1(r) + \phi_2(r)], \qquad (1.2)$$

where χ is the normalized wave function of the internal degree, and the probability density $P(r)$ of the particles falling on the screen where $r = R$ is given by the squared modulus $|\psi(R)|^2$ which contains the interference term $\phi_1(R)^* \phi_2(R)+$c.c. Now consider the case when an efficient detector such as an ultracold micromaser cavity with *no initial photons in it* is placed in both the paths. As the system passes through the cavities, it changes its internal state from χ to χ_- by emitting a single photon into one of the two cavities. Then the wave function of the correlated particle and the detectors is given by

$$\psi(r) = \frac{1}{\sqrt{2}} \left[\psi_1(r)\chi_-|1, 0\rangle + \psi_2(r)\chi_-|0, 1\rangle \right], \tag{1.3}$$

where $|1, 0\rangle$ and $|0, 1\rangle$ denote states of the detectors in which there is one photon in detector 1 and none in detector 2 and vice versa. Note that this is an 'entangled state' which is *not* the product of two factors, one referring to the system and the other to the detectors. The system and the detectors have got entangled by their interaction. In contrast to the previous case, the probability density of the particles at the screen is now given by

$$P(r) = \frac{1}{2} \left[|\psi_1|^2 + |\psi_2|^2 \right] \tag{1.4}$$

because $\langle 1, 0|0, 1\rangle = 0$ and the interference term $\psi_1^* \psi_2 \langle 1, 0|0, 1\rangle + \psi_2^* \psi_1 \langle 0, 1|1, 0\rangle$ vanishes. This is a consequence of the potential *information* contained in the detectors changing the outcome of the experiment, and *not due to uncontrollable alterations of the spatial wave function $\psi(r)$ resulting from the action of the detectors on the system under observation.* One might wonder whether the photon emission by the system might uncontrollably disturb the spatial wave function of the system. Careful calculations for centre-of-mass wave functions of atomic systems shows that this is not the case [5]. It is not even necessary actually to measure or observe the states of the detectors: their anti-correlation with the internal states of the system is sufficient for the loss of mutual coherence (see also Section 2.3.2). In fact, it is even possible in principle to erase the 'which path' information by letting the photon escape from the micromaser cavity. Such an arrangement is called a 'quantum eraser' [5] (see also Section 2.3.3). In this sense the micromaser cavities act as *Welcher Weg* ('which path') detectors which do not fall prey to the position–momentum uncertainty relations [6]. Of course, if the cavities initially contain classical microwave radiation with large (average) numbers of photons N_1 and N_2 with spreads $\sqrt{N_1}$ and $\sqrt{N_2}$, then the change in cavity 1 or 2 would be from $N_i \pm \sqrt{N_i}$ to $N_i + 1 \pm \sqrt{N_i}$ ($i = 1$ or 2). This change cannot be detected because $\sqrt{N_i} \gg 1$, so that in this case the cavity cannot contain

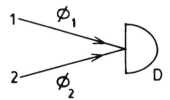

Fig. 1.3. Two independent sources of light 1 and 2 and a detector D (after Ref. [7]).

'which path' information and mutual coherence would not be lost. Now we will discuss a practical example of this in neutron interferometry

It has long been known that the interference produced by two light beams is related to their mutual coherence. We have just learnt that in quantum mechanics it is also related to the intrinsic indistinguishability of the particle paths. We will now show that there is a precise mathematical relation between these two features – the degree of indistinguishability of the paths equals the degree of coherence [7]. Consider Fig. 1.3 in which 1 and 2 are two sources of light and D is a detector. The quantum state of the light is represented by

$$|\psi\rangle = \alpha|1\rangle_1|0\rangle_2 + \beta|0\rangle_1|1\rangle_2 \tag{1.5}$$

with $|\alpha|^2 + |\beta|^2 = 1$. This is a coherent superposition of two states that are entangled : the photon can either originate in source 1 with probability $|\alpha|^2$ or in source 2 with probability $|\beta|^2$, but these two possibilities are indistinguishable. The corresponding density operator is given by

$$\begin{aligned}\rho_{\text{ID}} =\ & |\alpha|^2|1\rangle_1|0\rangle_2\ _2\langle0|_1\langle1| + |\beta|^2|0\rangle_1|1\rangle_2\ _2\langle1|_1\langle0| \\ & + \alpha\beta^*|1\rangle_1|0\rangle_2\ _2\langle1|_1\langle0| + \text{h.c.},\end{aligned} \tag{1.6}$$

where the subscript ID denotes indistinguishability. If, however, the density operator has the diagonal form

$$\rho_{\text{D}} = |\alpha|^2|1\rangle_1|0\rangle_2\ _2\langle0|_1\langle1| + |\beta|^2|0\rangle_1|1\rangle_2\ _2\langle1|_1\langle0|, \tag{1.7}$$

then it corresponds to an incoherent mixture of states. In this case it is possible in principle to distinguish the two possibilities, and the subscript D denotes this potential distinguishability.

Let us now consider an arbitrary one-photon state in the given Hilbert space with density operator

$$\begin{aligned}\rho =\ & \rho_{11}|1\rangle_1|0\rangle_2\ _2\langle0|_1\langle1| + \rho_{22}|0\rangle_1|1\rangle_2\ _2\langle1|_1\langle0| \\ & + (\rho_{12}|1\rangle_1|0\rangle_2\ _2\langle1|_1\langle0| + \text{c.c.}).\end{aligned} \tag{1.8}$$

It is easy to show that ρ can be uniquely decomposed as a linear sum of

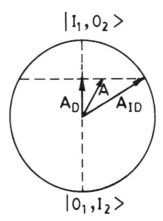

Fig. 1.4.   Illustration of the decomposition of ρ using the Bloch representation (after Ref. [7]).

ρ_{ID} and ρ_D:

$$\rho = P_{ID}\ \rho_{ID} + P_D\ \rho_D, \qquad (1.9)$$

where P_{ID} and P_D are probabilities for the sources to be indistinguishable and distinguishable respectively. This can be seen geometrically by representing the state on the unit sphere (Fig. 1.4). If we take the north pole to represent the state $|1_1, 0_2\rangle$ and the south pole to represent the state $|0_1, 1_2\rangle$, the pure state ψ (1.5) is represented by the unit vector $\mathbf{A}_{ID}$ while the mixed state ρ_D (1.7) is represented by the vertical vector $\mathbf{A}_D$. An arbitrary state ρ (1.8) represented by the vector $\mathbf{A}$ is obviously a linear sum of $\mathbf{A}_{ID}$ and $\mathbf{A}_D$ with coefficients that add up to unity.

Using (1.8) for the left hand side of (1.9) and (1.6) and (1.7) on the right hand side, we therefore get

$$\rho_{11} = |\alpha|^2, \qquad (1.10)$$
$$\rho_{22} = |\beta|^2, \qquad (1.11)$$
$$\rho_{12} = P_{ID}\,\alpha\beta^*, \qquad (1.12)$$

from which it follows that

$$\alpha\beta^* = [\rho_{11}\rho_{22}]^{1/2}\exp[i\arg\rho_{12}], \qquad (1.13)$$
$$P_{ID} = |\rho_{12}|/[\rho_{11}\rho_{22}]^{1/2}. \qquad (1.14)$$

Let us now consider the mutual coherence properties of the source field. If $E^{(+)}(r_1)$ and $E^{(+)}(r_2)$ are the positive-frequency parts of the electric field at r_1 and r_2, and

$$E^{(+)}(r_j) = Ka_j \quad (j = 1\,2) \qquad (1.15)$$

where K is some constant and a_j is the annihilation operator, then the second-order mutual coherence function $\Gamma_{12}^{(1,1)}$ for the sources is given by

$$\Gamma_{12}^{(1,1)} = \left\langle E^{(-)}(r_1) E^{(+)}(r_2) \right\rangle \tag{1.16}$$

$$= |K|^2 \, \mathrm{Tr}(a_1^{\dagger} a_2 \rho) = |K|^2 \rho_{21}, \tag{1.17}$$

and

$$\Gamma_{11}^{(1,1)} = |K|^2 \rho_{11} = |K|^2 |\alpha|^2, \tag{1.18}$$

$$\Gamma_{22}^{(1,1)} = |K|^2 \rho_{22} = |K|^2 |\beta|^2. \tag{1.19}$$

The normalized mutual coherence function is therefore given by

$$\gamma_{12}^{(1,1)} = \Gamma_{12}^{(1,1)} / \left(\Gamma_{11}^{(1,1)} \Gamma_{22}^{(1,1)} \right)^{1/2} = \rho_{21} / (\rho_{11} \rho_{22})^{1/2}. \tag{1.20}$$

Comparing this with (1.14), one obtains

$$|\gamma_{12}^{(1,1)}| = P_{\mathrm{ID}} \tag{1.21}$$

which shows that the degree of mutual coherence is equal to the degree of intrinsic indistinguishability of the two sources, showing that these two measures are equivalent and provide the fundamental link between the wave and particle domains in quantum theory.

Finally, let us see how this is related to the visibility of the interference pattern. The electric field $E^{(+)}$ at the detector due to the two sources is given by

$$E^{(+)} = a_1 \exp(i\phi_1) + a_2 \exp(i\phi_2), \tag{1.22}$$

where ϕ_1 and ϕ_2 are the phases due to the propagation from the source to the detector. Then the probability of detecting a photon at the detector is given by

$$\mathrm{Tr}\left[E^{(-)} E^{(+)} \rho \right] = \mathrm{Tr}\left[n_1 + n_2 + a_1^{\dagger} a_2 \exp i(\phi_2 - \phi_1) + \mathrm{h.c.} \right] \rho$$

$$= \rho_{11} + \rho_{22} + \rho_{21} \exp i(\phi_2 - \phi_1) + \mathrm{c.c.} \tag{1.23}$$

The third term exhibits the interference effect as $(\phi_2 - \phi_1)$ is varied. The visibility $\mathcal{V}$ of the fringe pattern is given by

$$\mathcal{V} = 2|\rho_{21}|/(\rho_{11} + \rho_{21}) = 2|\rho_{21}|$$

$$= 2|\gamma_{12}^{(1,1)}|(\rho_{11}\rho_{22})^{1/2}, \tag{1.24}$$

if use is made of (1.20). Now, there is a well-known inequality,

$$(\rho_{11}\rho_{22})^{1/2} \le \frac{1}{2}(\rho_{11} + \rho_{22}) = \frac{1}{2} \tag{1.25}$$

from which it follows that

$$\mathcal{V} \le |\gamma_{12}^{(1,1)}| = P_{\mathrm{ID}}, \tag{1.26}$$

with the equalities holding only when $\rho_{11} = \rho_{22}$. Thus, when the contributions of the two sources are equal, the degree of intrinsic indistinguishability of the two paths also equals the fringe visibility. Therefore, whenever the *entire* path of a photon from the source (aperture) to the detector becomes identifiable, the fringe pattern disappears *even in the absence of any physical disturbance of the interferometer* in the sense of the Heisenberg γ-ray microscope. This is built into the mathematical structure of quantum theory. The argument can be generalized and applied to two-photon interference experiments [8].

Recently, Englert [9] has derived the inequality $\mathscr{D}^2 + \mathscr{V}^2 \leq 1$ without requiring any use of the Heisenberg uncertainty relation, where $\mathscr{D}$ is a quantitative measure of path distinguishability introduced by him. According to this inequality the fringe visibility sets an absolute upper bound on the amount of 'which-path' information that is potentially stored in a detector.

These considerations show that the words 'wave' and 'particle' are divested of their absolute and objective classical meaning in quantum theory in which they retain only a *contextual* significance. It is possible to rephrase the concept of wave–particle complementarity in purely *quantitative* terms as the expression of the mutual exclusiveness of coherence and path distinguishability. It is, in fact, not necessary at all to use the words 'wave' and 'particle' which are borrowed from classical physics and are therefore of limited applicability in purely quantum mechanical situations. Nevertheless, one can, for convenience, also continue to use the words 'wave' and 'particle' consistently in the context of wave–particle complementarity provided they are used purely as *mnemonics* for 'mutual coherence'and 'path distinguishability' respectively.

1.3   Two-particle interferometry

Ever since Einstein, Podolsky and Rosen wrote their famous paper in 1935 and introduced the idea of entangled states, the Copenhagen school were compelled to change their stance vis-a-vis the inevitability of uncontrollably large disturbances accompanying the measurement of physical observables, because a measurement on one member of a pair of non-interacting particles in a quantum mechanically entangled state cannot in any mechanical way affect its partner instantaneously. This has extremely important implications for two-particle interferometry, that is, the interferometry of beams containing pairs of entangled states. The new field of two-particle interferometry has developed since the late 1980s and has already exhibited new non-classical phenomena, new confirmations of quantum mechanics and new violations of Bell's inequalities [10]. Here we will discuss only the complementarity of one-particle and two-particle interference. Fig. 1.5 shows a schematic two-particle four-beam interferometer using beam splitters H_1 and H_2. Let the two-particle entangled state be represented by

$$|\Psi\rangle = \frac{1}{\sqrt{2}}\left[|A\rangle_1|B\rangle_2 + |A'\rangle_1|B'\rangle_2\right]. \tag{1.27}$$

It is quite easy to show from this that the probabilities of joint and single detections by the various detectors are given by

$$P(U_1 U_2) = P(L_1 L_2) = \frac{1}{4}[1 - \cos(\phi_1 + \phi_2)], \tag{1.28}$$

$$P(U_1 L_2) = P(L_1 U_2) = \frac{1}{4}[1 + \cos(\phi_1 + \phi_2)], \tag{1.29}$$

$$P(U_1) = P(L_1) = P(U_2) = P(L_2) = \frac{1}{2}. \tag{1.30}$$

On the other hand, if the two-particle state is a product state of the form

$$|\Phi\rangle = \frac{1}{\sqrt{2}}\left(|A\rangle_1 + |A'\rangle_1\right)\frac{1}{\sqrt{2}}\left(|B\rangle_2 + |B'\rangle_2\right), \tag{1.31}$$

the probabilities of joint and single detections are given by

$$P(U_1 U_2) = P(U_1)P(U_2), \quad \text{etc.} \tag{1.32}$$

for all pairs of outcomes, and

$$P(U_1) = \frac{1}{2}(1 - \sin\phi_1), \tag{1.33}$$

$$P(L_1) = \frac{1}{2}(1 + \sin\phi_1), \tag{1.34}$$

$$P(U_2) = \frac{1}{2}(1 - \sin\phi_2), \tag{1.35}$$

$$P(L_2) = \frac{1}{2}(1 + \sin\phi_2). \tag{1.36}$$

It is clear from (1.28) and (1.29) that the entangled state $|\Psi\rangle$ represented by (1.27) produces sinusoidal variations with the phase angles or two-particle fringes, but no one-particle fringes. On the other hand, (1.32) through (1.36) show that the product state $|\Phi\rangle$ produces one-particle fringes but no genuine two-particle fringes, because the variation of $P(U_1 U_2)$ with ϕ_1 and ϕ_2 is only due to the separate variation of $P(U_1)$ and $P(U_2)$, as shown by (1.32).

These illustrate the quantum mechanical rules already emphasized. When the two particles 1 and 2 are prepared in the entangled state $|\Psi\rangle$, one can in principle determine which path particle 1 takes in going from S to H_1 by placing detectors in the beams B and B', and although this does

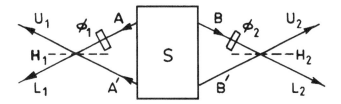

Fig. 1.5.   Schematic two-particle interferometer (after Ref. [10]).

not disturb particle 1, yet there is no one-particle interference; likewise for particle 2. It does not actually matter whether or not the detectors are in place. The very fact that a measurement 'could be performed without disturbing the system' in the spirit of Einstein, Podolsky and Rosen is sufficient to wash out the interference. However, once the pair of particles has gone beyond the two beam splitters H_1 and H_2 , there is no way to determine whether they have taken the composite path A, B or A', B', and so there is two-particle interference. If the particles are prepared in the product state $|\Phi\rangle$, there is no correlation of the one-particle paths, and so it is not possible to determine the path of one of the particles by observing the path that the other one takes. So the state $|\Psi\rangle$ exhibits maximum visibility of two-particle fringes and zero visibility of one-particle fringes, whereas the state $|\Phi\rangle$ exhibits exactly the opposite. It is possible to show for the most general state that can be formed from $|A\rangle_1, |A'\rangle_1, |B\rangle_2$ and $|B'\rangle_2$ that the visibilities in general satisfy the inequalities

$$0 \le \mathscr{V}_i \mathscr{V}_{12} \le \frac{1}{2} \quad (i = 1, 2) \tag{1.37}$$

which express complementarity of one and two-particle fringes.

A striking experimental confirmation of this result was obtained by Zou, Wang and Mandel in 1991 [11]. An outline of their experiment is shown in Fig. 1.6. Two similar nonlinear crystals NL1 and NL2 are pumped by two mutually coherent classical pump waves. Parametric down-conversion occurs at both the crystals, and a pair of *signal* and *idler* photons emerge from each. The paths of the signal photons s_1 and s_2 come together at the beam splitter BS_0, and the paths of the idler photons i_1 and i_2 can be either aligned or misaligned by shifting NL2. One can look for second-order interference between s_1 and s_2 in these two situations by varying their path difference slightly. Zou, Wang and Mandel observed second-order interference with the idler photons aligned even when the down conversions were spontaneous and the intensity of i_1 was weak so that an i_1 photon from NL1 and an s_2 photon from NL2 never accompanied each other. The interference disappeared when the i_1, i_2 connection was broken (by the insertion of a beam stop) or the two idlers were misaligned.

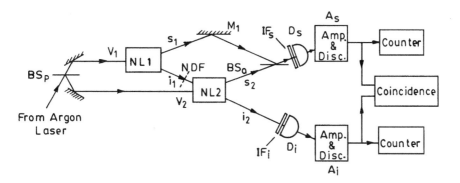

Fig. 1.6. Outline of the Zou–Wang–Mandel interference experiment (after Ref. [11]).

If the intensity of the i_1 field were very high, one would expect it to induce down-conversions in NL2, and i_1 and i_2 would be mutually coherent, and then s_1 and s_2 would also be mutually coherent. However, if the i_1 field is weak, this argument does not work, and it is difficult to understand in classical terms why s_1 and s_2 should still be coherent. The answer comes from quantum theory: i_1 induces coherence between s_1 and s_2 without inducing down-conversions in NL2. This can be shown by considering the density operator of the signal photons and tracing over the i_1 and 0 modes. The reason lies in the fact that with the idler photons perfectly aligned, the detectors D_s and D_i are unable to distinguish which crystal the detected photon comes from, and the corresponding two-photon probability amplitudes add. In fact, since signal and idler photons always accompany each other, every signal photon falling on D_s is accompanied by an idler photon falling on D_i, and D_i becomes even superfluous. However, once the i_1, i_2 connection is broken, it becomes possible in principle to place a 100 per cent efficient detector D_i and determine from the counts registered by it whether the signal comes from NL1 or NL2, and this potentiality is sufficient to destroy the interference recorded by D_s. It makes no difference whether or not D_i is even in place. *There is, therefore, in this case no question of a large uncontrollable physical disturbance acting on the system (in the sense of the Heisenberg γ-ray microscope) to wipe out the interference.* This is a most striking verification of the complementarity between mutual coherence (interference) and intrinsic indistinguishability of paths that quantum theory implies.

Ray and Home [12] have pointed out another experiment that involves γ-angular correlations on heavy-ion orbiting reactions which exhibits the same principle, namely that two states do not interfere if they are distinguishable *in principle*, even though they may not be distinguishable in practice.

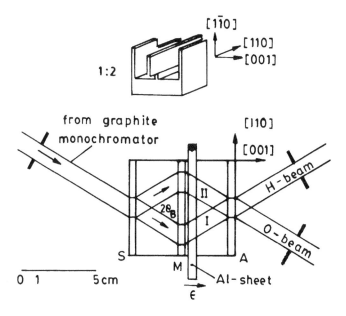

Fig. 1.7. Sketch of the perfect crystal neutron interferometer (after Ref. [13]).

1.4 Neutron self-interference experiments

Since 1974 neutron interferometry has offered a new access to a variety of research fields in physics [3], [13], [14]. It is an ideal tool to test the fundamental principles of quantum mechanics with massive particles on a macroscopic space–time scale. The interferometer consists of a perfect silicon crystal cut in the form of three plane-parallel plates on a common base (Fig. 1.7). The incident neutron beam is dynamically diffracted within the first plate and splits into two widely separated coherent beams. The middle plate superposes these two beams on the third plate which acts as an analyzer of the resulting interference pattern. Fig. 1.8 shows the plan view of the experiment. Another plane-parallel crystal plate (phase shifter) can be inserted into one of the beams, and a variable path difference introduced between the two beams by rotating it about a vertical axis. The flux of neutrons from even such high-flux reactors as the one in the Institute Laue–Langevin in Grenoble is so weak that no more than one neutron (whose wave packet is usually of the size of a normal postage stamp) passes through the apparatus at a time. In fact, when a neutron wave packet passes through the interferometer, its successor is still inside the uranium nucleus of the reactor fuel.

The state of the neutron after its dynamical diffraction by the first plate of the interferometer is represented by the wave function

$$\psi_0 = \psi_L + \psi_R \exp(i\phi) \tag{1.38}$$

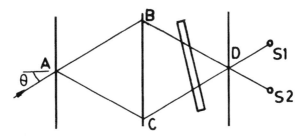

Fig. 1.8. Top view of the neutron interferometer experiment.

where L and R stand for the left and right beams respectively as seen from the top (plan view), and ϕ is the phase shift introduced by the phase shifter placed in the path of the right beam after the second plate. Rauch and Summhammer [14] carried out extremely interesting experiments by introducing an absorber in the path of the left beam. They used two different kinds of absorbers, a static (stochastic) absorber and a time-dependent (deterministic) absorber. Let us see what happens in these two cases.

1.4.1 Static (stochastic) absorber

The wave function ψ_0 is given by

$$\psi_0 = \sqrt{a}\,\psi_L + \psi_R \exp(i\chi), \tag{1.39}$$

where a is the absorption coefficient ($a < 1$). Then the neutron intensity (probability of detecting it) is given by

$$I_0 = |\psi_0|^2 = a\,|\psi_L|^2 + |\psi_R|^2 + 2\sqrt{a}\,\psi_L\,\psi_R\,\cos\chi. \tag{1.40}$$

The interference term is predicted to have an intensity proportional to $\sqrt{a}$.

1.4.2 Time-dependent (deterministic) absorber

In this case, the left neutron beam is chopped by a rotating toothed wheel, and every time the left beam is totally obstructed by the wheel, the right beam remains unobstructed (and we have therefore knowledge of the neutron's path), whereas at all other times both the beams propagate unattenuated and we have no knowledge of the neutron's path. The probability of the event is therefore the sum of the separate probabilities for each alternative. The speed of the chopper is so adjusted that on the average the left beam undergoes the same attenuation as in the static case. This time the intensity recorded by the same detector is therefore given by

$$\begin{aligned} I_0 &= a\,|\psi_L + \psi_R \exp(i\chi)|^2 + (1-a)\,|\psi_R|^2 \\ &= a\,|\psi_L|^2 + |\psi_R|^2 + 2a\,\psi_L\,\psi_R\,\cos\chi. \end{aligned} \tag{1.41}$$

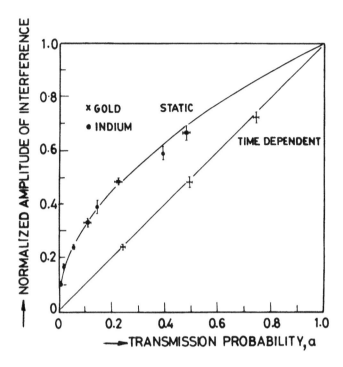

Fig. 1.9.   Observed amplitudes of the interference fringes for static and time-dependent absorbers (after Ref. [14]).

Notice that the intensity of the interference term is now predicted to be $2a$.

Rauch and Summhammer plotted the normalized intensities recorded by one of the detectors against the absorption coefficient in the static as well as the time-dependent cases and obtained the plots shown in Fig. 1.9. They are in agreement with the quantum mechanical predictions, although intuitively and classically it is hard to understand why the intensities should be different in the two cases when the average attenuation of one of the interfering beams is the same. What is also remarkable is that even when 99 per cent of the left beam was blocked, the interference pattern persisted with the same visibility!

It has been argued that one can generalize the complementarity idea and give it a quantitative form which would allow one to pass continuously from particle-like information to wave-like information (incompatible in Bohr's sense) by using information theoretical concepts [15], and that the above experiments with neutrons provide evidence of that [16]. Consider Fig. 1.10. Let the two coherent beams after the first plate be of amplitude a and let the amplitude of the left beam be b after absorption. Then the wave function of the neutron at the third plate or one of the detectors is

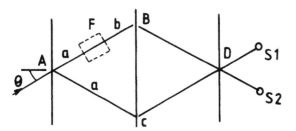

Fig. 1.10. Top view of a 'which path' experiment. The coherent beams are of amplitude a after the first plate; beyond the partial absorber the left beam amplitude is b (after Ref. [16]).

given by

$$\psi = [a\exp(ik_x x) + b\exp(i\phi)\exp(-ik_x x)]\exp(ik_z z) \qquad (1.42)$$

where k_x and k_z are the momenta of the two beams. We can choose a and b to be real, and the neutron intensity at the detector is given by

$$|\psi|^2 = a^2 + b^2 + 2ab\cos(2k_x x + \phi). \qquad (1.43)$$

One can now define a one-parameter measure of the wave nature W by

$$W = \frac{2ab}{a^2 + b^2} = \sin 2\beta \qquad (1.44)$$

by putting $a = R\cos\beta$ and $b = R\sin\beta$. When $\beta = \pi/4, b = R/\sqrt{2}$ and $W = 1$, which is a fully wave-like behaviour; we can have no knowledge of which path the particle takes. One can also define a particle-like measure P in the following way. Suppose there is no absorption ($b = a$). Then the probability of finding the neutron in the right beam is $1/2$. Therefore, with absorption present, a suitable particle measure would be

$$P = \frac{a^2/(a^2 + b^2) - \frac{1}{2}}{\frac{1}{2}} = \frac{a^2 - b^2}{a^2 + b^2} = \cos 2\beta. \qquad (1.45)$$

When there is complete absorption of the left beam, $b = 0$ and we know with certainty that the neutron must have taken the right path, and $P = 1$. It follows from (1.44) and (1.45) that

$$P^2 + W^2 = 1. \qquad (1.46)$$

This therefore provides a generalized quantitative expression for wave–particle complementarity, and one can pass from one extreme to the other by varying the single parameter β.

One might wonder whether the simultaneous presence of partially wave-like and partially particle-like properties implied by (1.46) constitutes a violation of Bohr's complementarity principle. The answer is, no. The reason is that once absorption occurs, the ensemble of neutrons is split into two incoherent parts, one that is absorbed and the rest. The ensemble of unabsorbed neutrons shows completely wave-like behaviour since their paths through the apparatus cannot be determined, whereas the ensemble of absorbed neutrons whose paths can be determined in principle behaves fully like particles. The concept of an entity that is neither fully a particle nor fully a wave is essentially non-classical, and is not encompassed by Bohr's wave–particle complementarity [17].

Let us look at another interesting experiment in neutron interferometry that was suggested by Vigier and co-workers and performed by Badurek, Rauch and Tuppinger [18]. It raises interesting questions regarding the nature of wave–particle duality ('particles *and* waves' favoured by de Broglie versus 'particles *or* waves' favoured by Bohr and his Copenhagen school) and the measurement problem. The arrangement is shown schematically in Fig. 1.11. The original symmetric triple-crystal interferometer is replaced by a skew-symmetrically cut interferometer as shown, so that the two coherent sub-beams inside the interferometer propagate well separated and in parallel directions over a distance of several centimetres. This facilitates the installation of devices which act on one of the sub-beams only without affecting the other. A chopper cuts the incident neutron beam into wave packets whose dimensions are smaller than those of two separate radio-frequency (r.f.) resonance spin-flip coils I and II that are inserted into the paths of the two sub-beams to invert the spin state of each sub-beam. The coils are operated alternately and exclusively when each corresponding sub-beam passes through it. The spin reversal takes place when every neutron exchanges a photon of energy $\hbar\omega_r$ with one of the two coils, *but we do not know with which one*. Afterwards, the two coherent sub-beams recombine in the third slab to build up an interference pattern. From a *realist* de Broglian point of view, Vigier and his co-workers claim that because of the indivisibility of each exchanged quantum or photon, the energy transfer has to be associated with the particle properties of the neutron. They conclude therefore that every neutron follows a definite *trajectory* inside the interferometer (its particle property) while the wave extends over both paths simultaneously but is invisible when the particle is detected. There exists the possibility to deviate one of the sub-wave packets in the second slab (*delayed choice*) once the neutron has passed through coil I. This makes more precise which path the neutron has chosen.

However, the actual experimental results obtained by Badurek *et al.* are in full agreement with the predictions of quantum mechanics and are identical for both interpretations ('particles *and* waves' and 'particles *or* waves'). Nevertheless, well-defined particle properties can also be attributed to the neutron such as its mass, its spin and the associated magnetic moment, its

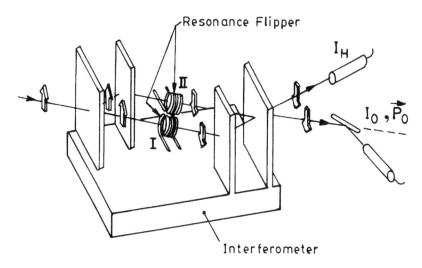

Fig. 1.11. The Rauch–Vigier experiment with a skew-symmetric neutron inter-ferometer (after Ref. [18]).

effective-mass radius and its internal structure consisting of one 'up' and two 'down' quarks. Let us now see what quantum mechanics predicts for this double resonance experiment which can be done with polarized as well as unpolarized neutrons. Let us first consider the incoming neutrons to be polarized in the $\pm z$-direction. So the initial wave function is

$$\psi \propto \exp(-iEt/\hbar)|\pm z\rangle. \tag{1.47}$$

The spin-flip operator U of an rf coil transforms ψ into the final state

$$\psi' = U\psi \propto \exp\left[-\frac{i}{\hbar}(E \mp \hbar\omega_r)t\right]|\mp z\rangle. \tag{1.48}$$

Denoting the operators of the two rf flippers in the beam paths L and R by U_L and U_R and the phase shift due to the phase shifter plate by χ, one can calculate the final wave function at the detector in the forward direction and hence the neutron intensity there as well as the polarization:

$$\psi' = \left(U_L + e^{i\chi}U_R\right)\psi. \tag{1.49}$$

To begin with, let us assume that the incident neutrons are completely polarized in the z direction $(+P_z)$, and that the two spin flippers are operating in the synchronous mode, i.e., their effective lengths are equal, their resonance frequencies are identical and their initial oscillation phases are also the same. Then

$$\begin{aligned}\psi' \propto \;& \exp\left[-i(E/\hbar - \omega_r)t\right]|-z\rangle \\ & + e^{i\chi}\exp\left[-i(E/\hbar - \omega_r)t\right]|-z\rangle,\end{aligned} \tag{1.50}$$

$$I \quad \propto \quad 1 + \cos \chi, \tag{1.51}$$
$$P \quad = \quad -p_z. \tag{1.52}$$

Thus, although the spin flips and the final polarization is oriented opposite to the initial one, the interference does not disappear.

Analogous results are obtained for synchronous operation with a distinct but constant phase difference between the two oscillating fields, but with asynchronous operation the interference disappears because of the randomly fluctuating phase difference. If the incident beams are unpolarized, then one obtains the same interference effect but no final polarization in the synchronous mode, and no interference for the asynchronous operation for the same reason. The result for the synchronous operation with a constant phase difference is a little more complicated.

The experimental plots are shown in Fig. 1.12. They are in complete agreement with the predictions of quantum mechanics. However, they raise interesting issues of interpretation. Does the energy transfer within the resonance coil trigger a collapse of the wave function and constitute a measurement process? The fact that the interference is not wiped out indicates that it does not. As we have seen before, since the r.f. coils contain classical radiation, they cannot act as *Welcher Weg* detectors. However, only quanta $\hbar\omega_r$ within a narrow frequency band are excited within the r.f. coils, which suggests that the energy transfer to an individual neutron takes place in a single coil, and so the neutrons must follow definite paths through the apparatus although the persistence of the interference pattern indicates that these paths cannot be distinguished even in principle!

Let us finally look at another intriguing experiment. If one of the neutron beams inside the silicon-crystal interferometer is allowed to traverse a region of potential V extending over a variable distance D (achieved by inserting a polished pure aluminium slab of variable thickness in the path), the potential has the effect of increasing the transit time of the neutron wave packet in that path relative to the one in the other path. If the relative spatial shift between the two wave packets is large enough, they will not overlap on the third plate in the interferometer, and the contrast in the interference pattern should disappear. This is indeed what is observed: there is a substantial and continuous loss of contrast as the thickness of the aluminium slab is increased [19]. This can be interpreted as being due to the finite length of the neutron wave packets. The most surprising thing that happens if a momentum filter is now inserted between the interferometer and the detector is that a marked modulation appears in momentum space [20]! This is obviously due to the fact that the constituent momentum states forming the neutron wave packets are infinitely extended and so overlap and interfere! The fact that the wave packets themselves do not overlap spatially and are yet able to interfere

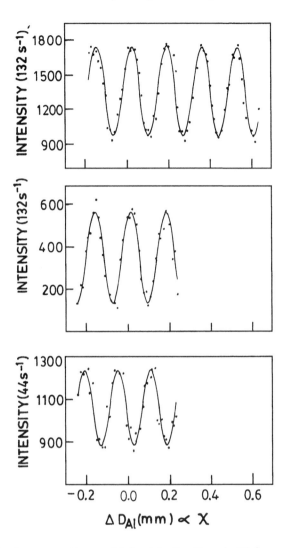

Fig. 1.12. Interference patterns obtained in the Rauch–Vigier experiment (after Ref. [18]).

and influence each other is significant for an understanding of nonlocal EPR correlations which we will have occasion to discuss at length later in Chapter 9. Suffice it to say for the moment that this experiment offers a possible understanding of EPR nonlocality as the far-reaching action of the plane wave components of fairly well localized wave functions. The experiment also showed that it is possible to generate squeezed states in neutron optics.

A curious example of interference among wave functions of a γ-ray emitting source (^{12}C) co-existing with 'which path' information has been pointed out by Ray and Home [21]. This coexistence is, however, consistent with the formalism of quantum mechanics because the interference in question results from the coherence of the nuclear angular momentum states whereas the 'which path' information available is in configuration space and can be obtained without involving any entanglement with the interfering states.

1.5   Electron self-interference

Tonomura's team at the Hitachi Laboratory in Japan have developed remarkable tests of the wave–particle duality of electrons by developing coherent electron beams of very low intensity [22]. The electronic analog of the double-slit experiment uses a special electron gun that emits coherent wave fronts whose central parts are blocked to create a 'double slit'. The unobstructed parts of the wave fronts can be deflected inwards by electric fields as they pass through these slits, and they eventually recombine and fall on a two-dimensional array of detectors. Every wave front produces a single spot , i.e., it is detected by a single detector, and the interference pattern builds up over a long exposure. The theoretical analysis is the same as for neutron interferometry. The wave aspect of electrons was also dramatically demonstrated through the development of electronic holography and a clean test of the Aharonov–Bohm effect.

Before passing over to self-interference experiments with single-photon states, let us note that causal explanations of the observed electron and neutron self-interference effects can be given in terms of well-defined particle trajectories and the quantum potential[23], which shows that the standard Copenhagen interpretation of quantum mechanics and its variants are derived from philosophical assumptions that are not necessarily implied by the mathematical formalism and success of quantum mechanics.

Selleri and his group in Bari who are physical *realists* like de Broglie, have been exploring the implications of 'empty waves' or 'ghost waves' as applied to massive particles [24]. According to them, these waves can induce quantum transitions in atoms but cannot fire detectors. This explains all negative coincidence tests, because when the particle aspect chooses one path only, the empty wave in the other path cannot fire the detector. According to Vigier, however, who is also a realist in the de Broglian sense, the wave in the duo composed of the particle and the wave carries energy and is not empty, and can explain energy variations along the trajectories as well as astrophysical red-shifts [3]. A recent experiment by Mandel [25], claims to rule out empty waves of the type proposed by Selleri though not a variant of it implied by the de Broglie–Bohm theory [26] which we will discuss later.

1.6 Single-photon states and complementarity

Finally, let us turn to wave–particle duality of single-photon states. For such states unambiguous evidence of complementarity between interference and 'which path' information had to wait until technological advances made it possible to produce and detect them. The first experiment of this kind was performed by Aspect, Grangier and Roger around 1986/87 [27], [28], [29]. A number of experiments had been done earlier since 1909 [30], to demonstrate interference with highly attenuated pulses of quasi-monochromatic light of frequency v whose energies were so low they that it was believed they could not contain more than one photon of the same frequency and whose flux was so low that no more than one pulse could pass through the interferometer at a time. Each pulse produced a single spot on the photographic plate, and it took about six months of exposure time for Taylor to develop the fringe pattern in 1909. It was argued that these experiments demonstrated the wave–particle duality of light, the particle aspect manifesting itself in the discrete spots and the wave aspect in the interference fringes. However, clearly there was no 'which path' information in these spots and there was no arrangement in any of these experiments to demonstrate the mutual incompatibility between the fringe pattern and 'which path' information that lies at the heart of the Einstein–Bohr debate and complementarity. This is exactly what Aspect and his collaborators provided, and in the process also demonstrated that it was impossible to get 'which path' information with the kind of low-flux weak pulses that had been used earlier – one needs to use a low flux of genuine single-photon pulses that were not available earlier.

Aspect *et al.* pointed out another important feature of these experiments that is often not appreciated. And it is this: although the discreteness of the spots (photodetection) can be explained if one accepts the existence of photons, *they can also be attributed to the quantization of the detector alone*. There is therefore *no logical necessity to introduce the concept of the photon to describe the interference of weak light*. Let us see how.

Let us consider a simple model in which the detector is an atom with a ground state $|g\rangle$ and a continuum of excited states $|e\rangle$, with a gap W_T (Fig. 1.13). A *classical* electromagnetic field is incident on the atom and interacts with it. Let the interaction Hamiltonian be $\vec{E}.\hat{\vec{D}}$ where the electric field $\vec{E}$ is a number and $\hat{\vec{D}}$ is the atomic electric dipole operator. If $\vec{E} = \vec{E}_0 \exp(i\omega t)$, the transition rate from the state $|g\rangle$ to the state $|e\rangle$ is given by Fermi's Golden Rule:

$$\frac{d}{dt}\mathscr{P}_{g\to e} = \frac{\pi}{2\hbar} |\langle e|\hat{\vec{D}}|g\rangle|^2 \, \mathscr{E}_0^2 \rho(e)\delta(E_e - E_g - \hbar\omega). \qquad (1.53)$$

This contains all the features of the photoelectric effect. The existence of

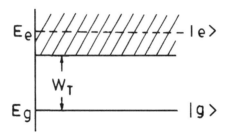

Fig. 1.13.   Model of a detector for the photoelectric effect (after Ref. [28]).

a threshold is reflected in the feature that the density of excited states $\rho(e)$ vanishes if $E_e - E_g < W_T$. The final energy of the system is $E_e = E_g + \hbar\omega$, and therefore the kinetic energy of the emitted electron is $\hbar\omega - W_T$. Also, the probability of photodetection is proportional to the intensity $\mathscr{E}_0^2$. Thus, *Einstein's interpretation of the photoelectric effect by no means implies the necessity of describing light as composed of photons.* Indeed, there were many stalwarts like Planck, Nernst, Warburg and Bohr who refused to take Einstein's light-quantum hypothesis seriously for a long time [31]. It is only now that the truly quantum nature of light is beginning to be observed definitively through the production of special states of light like single-photon states and squeezed states [32] that the light-quantum hypothesis has been convincingly vindicated.

The strategy followed by Aspect *et al.* was to use anti-correlation or anti-coincidence on a beam splitter to obtain 'which path' information. The quantum theory of light predicts perfect anti-correlation for singles detections on the two sides of a beam splitter, whereas a classical light pulse, however weak, would be divided on a beam splitter and therefore produce a minimum degree of correlated or coincident counts. It is possible to make these considerations more quantitative. Let a source of light pulses impinge on a beam splitter (Fig. 1.14). A triggering system produces gates of duration T, synchronized with the light pulses. The detections are authorized only during the gates. A coincidence is counted when both the photomultiplers PM_r and PM_t register a detection during the *same* gate. Let N_1 denote the rate of gates, N_r and N_t the rates of singles detections by PM_r and PM_t respectively and N_c the rate of coincidences. Then the probabilities for a single count during a gate are given by

$$P_t \;=\; N_t/N_1 = \beta_t\,\eta\,T\,\langle i_n \rangle, \tag{1.54}$$

$$P_r \;=\; N_r/N_1 = \beta_r\,\eta\,T\,\langle i_n \rangle, \tag{1.55}$$

and that for a coincident count by

$$P_c = N_c/N_1, \tag{1.56}$$

where

$$i_n = \frac{1}{T} \int_{t_n}^{t_n+T} I(t) \, dt \qquad (1.57)$$

is the average intensity during the nth gate and the brackets $\langle \rangle$ denote an average over the ensemble of gates; β_r and β_t are reflection and transmission coefficients respectively of the beam splitter and η is the detection efficiency of the photodetectors. When the singles probabilities are small compared to unity, one can write

$$P_c \approx \beta_r \beta_t \eta^2 \, T^2 \, \langle i_n^2 \rangle. \qquad (1.58)$$

Now, the standard Cauchy–Schwartz inequality

$$\langle i_n^2 \rangle \geq \langle i_n \rangle^2 \qquad (1.59)$$

holds for the bracket average. Therefore, for any classical wave description of the experiment of Fig. 1.14, one expects

$$P_c \geq P_r P_t, \qquad (1.60)$$

or, equivalently

$$\alpha \geq 1, \quad \alpha = \frac{P_c}{P_r P_t}. \qquad (1.61)$$

This result states that there is a *minimum* rate of coincidences counts, corresponding to 'accidental coincidences', for a *classical* wave divided on the beam splitter. On the other hand, $\alpha = 0$ for a single-photon state because the state is

$$|\psi\rangle = \frac{1}{\sqrt{2}} \left[|1,0\rangle |D_1\rangle + i |0,1\rangle |D_2\rangle \right] \qquad (1.62)$$

after the beam splitter, $|D_1\rangle$ and $|D_2\rangle$ being the detector states, and the probability of joint detection vanishes because of the orthogonality of the states $|1,0\rangle$ and $|0,1\rangle$. This therefore provides an empirical or operational criterion characterizing single-particle behaviour of light – the violation of the inequality (1.61) would indicate that the light pulses cannot be described as wave packets that divide on the beam splitter but rather as single photons that cannot be so divided and detected simultaneously on both sides of the beam splitter.

When Aspect *et al.* used light from a pulsed photodiode that was attenuated to a level corresponding to one detection per 1000 emitted pulses with a detector quantum efficiency of about 10 per cent, the average energy per pulse was estimated to be about 0.01 photon. Such a source would certainly have been considered a source of single photons. Nevertheless, the quantity α was consistently found to be equal to unity, i.e., no anti-correlation was observed. This showed that light emitted by an

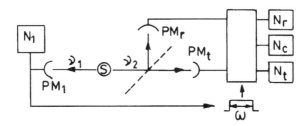

Fig. 1.14. Study of detection correlations after a beam splitter. The detection of the first photon of the cascade produces a gate ω, during which the photomultipliers PM_t and PM_r are active (after Ref. [27]).

attenuated classical source does not exhibit single-photon behaviour, and one cannot get 'which path' information using such a source. A fairly localized wave packet with a small enough energy does not necessarily imply that it would behave like a 'particle' and not divide itself on a beam splitter. Indeed, such wave packets can obviously be built up by superposing classical waves. It is therefore impossible to demonstrate the complementarity between interference and 'which path' information using a weak classical source.

Although an excited atom emits a single photon, many atoms of a classical source are 'in view' of the detectors, and the number of excited atoms fluctuates. Consequently, the emitted light can be described by a density matrix reflecting these fluctuations. For a Poissonian fluctuation of the number of emitting atoms, one can show that the statistical properties of the light cannot be distinguished from those of classical light.

In order to observe non-classical features of light, it is therefore necessary to isolate single-atom emissions. This was achieved by Aspect *et al.* using the two-photon radiative cascades from a calcium source (Fig. 1.15). The atoms were excited by two-photon absorption from stabilized continuous wave (cw) lasers to an upper level which then de-excited in two successive steps, emitting two photons at different frequencies v_1 and v_2. The intermediate level was metastable and had a life-time $\tau_s = 4.7$ ns. By choosing the rate of excitation of the cascades such that $N \ll \tau_s^{-1}$, they achieved cascades well separated in time. They used the detection of the first photon v_1 as a trigger for a gate of duration $T \approx 2\tau_s$. During a gate the probability of a photon v_2 coming from the *same* atom that emitted v_1 was much larger than that from a *different* atom in the source. They were therefore in a situation close to an ideal single-photon source.

The quantum mechanical prediction for α is

$$\alpha_{QM} = \frac{2f(T)NT + (NT)^2}{[f(T) + NT]^2}, \tag{1.63}$$

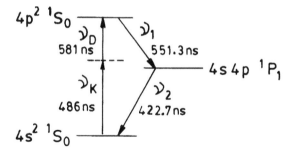

Fig. 1.15. Two-photon radiative cascade of calcium used to generate single-photon wave packets (after Ref. [28]).

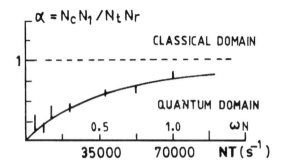

Fig. 1.16. The anti-correlation parameter α as a function of ωN (the number of cascades excited during a gate) and the trigger rate NT (after Ref. [27]).

where $f(T)$ is a quantity very close to unity in the experiment, being the product of the factor $[1 - \exp(-T/\tau_s)]$ (the overlap between the gate and the exponential decay) and a factor somewhat greater than unity related to the angular correlation between v_1 and v_2 [33]. The smaller NT is compared to $f(T)$, the stronger will be the anti-correlation ($\alpha \ll 1$). The results are shown in Fig. 1.16 which shows a plot of α as a function of NT. It is clear that the violation of the inequality (1.61) increases as NT decreases. The maximum violation of more than 13 standard deviations was obtained for a counting time of five hours. The value of α was 0.18 ± 0.06, corresponding to a total number 9 of coincidences instead of a minimum of 50 coincidences expected for a classical model of light. This was characteristic of single photons as predicted by quantum theory, and the anti-correlated counts on the two sides of the beam splitter gave 'which path' information. With such a source available, Aspect *et al.* could turn to a study of single-photon interference.

They constructed a Mach–Zehnder interferometer in which the two detectors were removed and placed behind a second beam splitter on

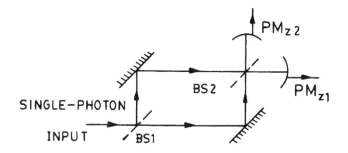

Fig. 1.17.   The Mach–Zehnder interferometer (after Ref. [27]).

which the two beams were recombined with the help of two mirrors, as shown in Fig. 1.17. The detectors were gated as in the previous experiment and once again their counts were found to be anti-correlated, but this time they did not contain any 'which path' information which was lost after the second beam splitter. When the path difference between the two beams was varied with the help of a phase shifter, the probabilities of single-photon detection by each detector varied sinusoidally. The fringe visibility was close to unity. This completed the demonstration of complementarity between 'which path' information and interference.

It is worth noting that the question of incompatible descriptions of these two mutually exclusive experiments arises only if one insists on using *classical* concepts like particles or waves. The mathematical formalism of quantum mechanics, on the other hand, provides a unique description of light in both the experiments in terms of the same 'state vector' (or density matrix). It is the observed quantity which is *contextual* and changes according to the experimental arrangement, not the abstract and mathematical description of light.

1.7   The double-prism experiment

We have seen that the degree of mutual coherence of two beams depends on their intrinsic path indistinguishability, and even the mere possibility of obtaining 'which path' information is sufficient to destroy this coherence. This is the basis of the complementarity in the sense of *mutual exclusiveness* of the wave and particle descriptions in quantum theory. *According to Bohr this mutual exclusiveness also extends to the classical concepts of waves and particles.* He presumably arrived at this view point from (a) his insistence on the use of common language, suitably refined by classical physics, for the unambiguous communication of every experimental result and (b) the assumption that interference is 'the only means of defining the concepts of frequency and wavelength of a photon' [34]. The question arises as to whether Bohr's extension of complementarity to the classical concepts of waves and particles holds in *every situation*. A little reflection shows that it does *not*, because Bohr's assumption (b) is not true. There are characteristic features of wave-like propagation like tunneling and birefringence of single-photon states in which the *interference* of the alternative amplitudes (for the reflected and transmitted rays in tunneling and for the ordinary and extraordinary rays in birefringence) plays no pertinent role, and yet these phenomena can just as well be used to define and measure wavelengths and frequencies as interference. For example, tunneling occurs only when the tunneling gap is less than the wavelength of the incident photons and can, therefore, be used to measure their wavelengths, and birefringence is the result of dispersion in the refracting medium. *There is, however, nothing in the mathematical structure of quantum mechanics that rules out simultaneous path distinguishability and tunneling or simultaneous path distinguishability and birefringence.* If the concept of a 'wave' in quantum theory is extended to include tunneling and birefringence in addition to coherence, as in classical physics, it follows that quantum mechanics does not rule out the possibility of observing wave- and particle-like behaviours *at the same space–time point* (the point of incidence of a photon on the boundary between two media). The classical concepts of waves and particles cannot obviously be consistently applied to such situations. We will now discuss an actual experiment of this kind proposed by Ghose, Home and Agarwal [35].

The classical analogue of the experiment was done by J. C. Bose way back in 1897 as reported in Sommerfeld's *Optics* [36]. A beam of microwaves directed at a 45° asphalt prism was totally internally reflected by the 45° face. When a second similar prism was placed face to face in contact with the first prism, the beam passed straight through. However, when the gap between the prism faces was increased but kept significantly smaller than the wavelength of the beam, a portion of the beam was internally reflected and the rest tunneled across the gap. This was a striking confirmation of the wave nature of the radiation with which Bose was experimenting. A similar experiment can also be done with visible light, only in this case the gap between the prism faces must be controlled and kept at about several tens of nanometers.

Ghose, Home and Agarwal pointed out that if this experiment was performed with 'single-photon states' of light, the interpretation of the results in terms of classical wave and particle pictures would acquire a new significance. Should the photon behave like a classical wave, a part of it must be reflected and at the same time the rest of it must tunnel across the gap. On the other hand, should the photon behave strictly like a classical particle, it should not be able to tunnel across the gap because tunneling is a purely wave phenomenon. Quantum optics, however, predicts that a photon will be either reflected by or tunnel across the gap, but *not at the same time*. Thus, a photon will reveal itself as a wave when it tunnels across the gap, and *at the same time* as a particle because it is indivisible and follows a path. Thus *both wave and particle behaviours of light should be observable simultaneously.*

Consider Fig. 1.18. In classical optics the field amplitudes a, c and d obey the relations $d = \gamma a, c = \alpha a$, where γ and α are respectively the reflection and transmission amplitudes. For certain angles of incidence, total internal reflection occurs and the waves in region I are *evanescent*. (These evanescent waves propagate parallel to the surface of the first prism, and though the wave extends into the gap and dies off exponentially, the Poynting vector parallel to the initial direction of propagation vanishes.) If the thickness h of this region is large compared to the wavelength λ then, by the time the field reaches the surface of the second prism, the amplitude has decayed exponentially to almost zero and no transmission takes place. In quantum optics, the quantities c, d and a must be treated as annihilation operators. Moreover, in order to maintain the commutation relations, one has to add the vacuum field b at the open port. Thus we have

$$\begin{aligned} c &= \alpha a + \beta b, \\ d &= \gamma a + \delta b, \end{aligned} \tag{1.64}$$

and the commutation relations

$$\begin{aligned} \left[a, a^\dagger \right] &= \left[b, b^\dagger \right] = 1, \\ \left[a, b^\dagger \right] &= 0, \\ \left[c, c^\dagger \right] &= \left[d, d^\dagger \right] = 1. \end{aligned} \tag{1.65}$$

Note that $|\alpha|^2 + |\gamma|^2$, since the prisms are supposed to be lossless. Moreover, β is related to γ through at most a phase factor. The probability $p_d(1)(p_c(1))$ of detecting a photon at the detector $D_1(D_2)$ is given by

$$p_d(1) = \text{Tr} \left[\rho | 1 \rangle_{dd} \langle 1 | \right], \tag{1.66}$$

where $| 1 \rangle_d$ is the single-photon state associated with the mode d. Assuming the input states to be $| 1 \rangle_a | 0 \rangle_b$, these probabilities can be calculated to be

$$\begin{aligned} p_d(1) &= |\gamma|^2, \\ p_c(1) &= |\alpha|^2. \end{aligned} \tag{1.67}$$

Note that these results are the same as one would get on the basis of classical electrodynamics. Thus tunneling would occur as long as it occurs in classical theory. In order to see the quantum features, let us compute the joint probability $p_{cd}(1, 1)$ of detecting one photon at D_1 and one

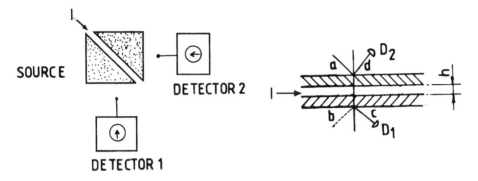

Fig. 1.18. Principle of the double-prism experiment (after Ref. [34]).

photon at D_2:

$$p_{cd}(1, 1) = \text{Tr} \left[\rho \,|1\rangle_c \,|1\rangle_{dd}\langle 1|_c \langle 1| \right]. \tag{1.68}$$

Using (1.65) this reduces to

$$p_{cd}(1, 1) = o, \tag{1.69}$$

which implies that the counts in the two detectors are anti-correlated. In other words, the double prism behaves like the beam splitter in the experiment of Aspect *et al. with the additional feature that the transmission through it can only occur by tunneling, which is a wave phenomenon.*

If the light source is semi-classical, the second-order correlation function is unity. Therefore, with equal probabilities of transmission and reflection, one has

$$\frac{\langle n_1 n_2 \rangle}{\langle n \rangle} = \frac{1}{4} \langle n \rangle, \tag{1.70}$$

where $n = n_1 + n_2$ is the total number of incident photons in a state, and $\langle \rangle$ denotes the ensemble average [32]. Therefore the ratio of coincidences to singles is proportional to the singles counts. On the other hand, as we have seen, if one uses single-photon states, the second-order correlation vanishes.

Mizobuchi and Ohtaké [37] performed the experiment. They used a pulsed laser diode to measure the coincidence counts (Fig. 1.19), and reduced the light intensity with the help of neutral density filters to as low as 10^4 photons per second. With an apparatus size of less than one metre, this ensured that there was never more than one photon in the apparatus at any given instant. Yet, the measured ratio of coincidences to singles counts was found to be nearly proportional to the singles counts (Fig. 1.20), as given by (1.70). Thus, no non-classical wave-like or single particle-like behaviour was observed with semi-classical light, as in the experiment of Aspect *et al.*

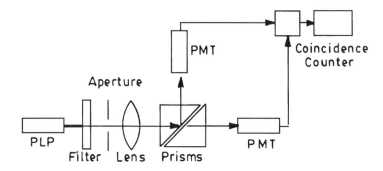

Fig. 1.19.   The coincidence experiment (after Ref. [37]).

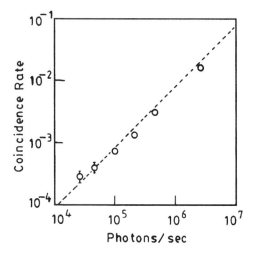

Fig. 1.20.   Coincidences as a function of singles counts per second (after Ref. [37]).

To do the anti-coincidence experiment (Fig. 1.21) proposed by Ghose, Home and Agarwal, they used the parametric down-conversion technique [38] to produce correlated photon pairs. One of the photons served as the single-photon source. The advantage of the down-conversion technique over the atomic cascades is that the former is more controllable and gives higher gain because the down-converted photons strictly satisfy the momentum conservation law. They used the third harmonic of a pulsed Nd:YAG laser which has a typical wavelength of 355 nm, and the beam was injected on a BBO crystal in which the down-converted photon pairs of wavelength 710 nm were generated. The intensity of the light was again reduced with the help of neutral density filters to the single-photon intensity level before it was incident on the prism. The gap between the prism faces was controlled by putting Langmuir–Blodgett

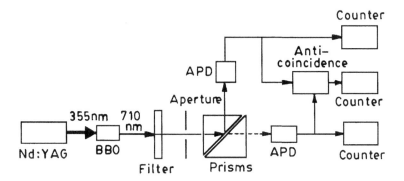

Fig. 1.21.   The anti-coincidence experiment (after Ref. [37]).

films between the prisms, leaving the passage of the light untouched. The relation between tunneling and reflection was measured as the gap width was varied. It was found that as the gap decreased, tunneling increased (Fig. 1.22), showing the classical wave nature of the light. The reflected and transmitted light were detected by avalanche photodiode (APD) single-photon detectors with an efficiency of 38 per cent. The resolving time of anti-coincidence was determined by the input rectangular pulse duration of 600 ns. The counter units were gated on for 20 µs synchronized with each laser pulse to minimize the dark counts. The results are shown in Fig. 1.23 in which the ratio of anti-coincidence counts to the number of signal counts from either the reflected or the transmitted light is plotted against the detected signal counts per second at each detector. This ratio should be unity if there is complete anti-coincidence. The experimental results tend to show that this is the case. The uncertainties arise from the lack of long-time stability of the Nd:YAG laser used and the consequent lack of statistical accuracy. Nevertheless Mizobuchi and Ohtaké concluded:

> Therefore in the single-photon states, light shows its indivisibil-
> ity – manifestation of the classical particle property, while still
> being able to tunnel across the gap between the prisms – man-
> ifestation of the classical wave property, at the same time....
> In summary, our anti-coincidence experiment supported the
> prediction of quantum optics, namely, light showed both the
> classical wave-like and the classical particle-like pictures simul-
> taneously. This is in contrast to the conventional interpretation
> of the duality principle.

Unnikrishnan and Murthy [39] have made a systematic error analysis of this experiment and shown that although the observed anti-coincidence rate is smaller than the classical prediction, it implies a larger coin-cidence rate than the classical prediction because of the large inten-

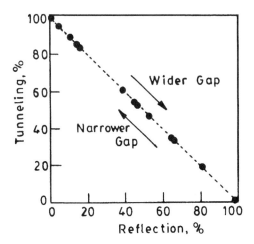

Fig. 1.22. The relation between reflection and tunneling as the gap between the prisms is varied (after Ref. [37]).

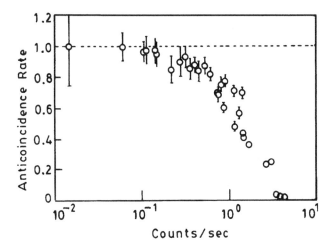

Fig. 1.23. The anti-coincidence rate plotted against signal counts per second (after Ref. [37]).

sity fluctuations produced by the Nd:YAG laser. They suggest that the use of a relatively more stable cw laser for the down-conversion would be preferable as the intensity fluctuations would be smaller and the count rates orders of magnitude higher. Such an experiment is in progress.

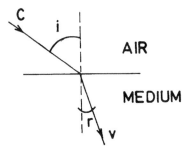

Fig. 1.24. Principle of the single-photon birefringence experiment.

1.8 Birefringence and complementarity

Refraction is yet another phenomenon which conclusively established the wave nature of classical light. Any corpuscular classical theory of light, like that of Newton, implies that light must travel faster in an optically denser medium than in vacuum in order to be consistent with the observed law of refraction (Snell's law). This follows if one makes use of the condition that the transverse component of the velocity must be continuous across the boundary between the two media (Fig. 1.24):

$$c \sin i = v \sin r \qquad (1.71)$$

where i and r are the angles of incidence and refraction ($i > r$) and c and v are the velocities of light in vacuum and the optically denser medium respectively. Then, the refractive index μ is given by

$$\mu = \frac{\sin i}{\sin r} = \frac{v}{c} > 1. \qquad (1.72)$$

The classical electromagnetic theory of light, on the other hand, implies the opposite, namely

$$\mu = \frac{\sin i}{\sin r} = \frac{c}{v} > 1. \qquad (1.73)$$

Fizeau's experiment [40] on the speed of light in water conclusively showed that $v < c$, ruling out any classical particle theory of light.

There are birefringent crystals in which the incident light is split into two beams, the ordinary and the extra-ordinary rays, both of which have refractive indices larger than unity and therefore propagate as waves. If a single-photon state is incident on such a crystal, it will be in a superposition of the two possible paths (corresponding to the ordinary and extra-ordinary rays). If two detectors are placed in these two paths, their counts will be anti-correlated, giving us 'which path' (or particle-like) information. Yet, the very fact that the refractive indices for both the paths are greater than unity can only be explained in terms of wave-like propagation. Here again Bohr's extension of complementarity to the classical concepts of waves and particles fails. An experiment was carried out to test this prediction. Although the statistical accuracy was somewhat poor, the results tend to support the prediction [39]. A more accurate experiment using a cw laser is in progress.

1.9   Conclusion and summary

We therefore conclude that although the complementarity principle holds in every situation in which non-commuting observables are involved, in the case of wave–particle duality Bohr's extension of the principle to the *classical* concepts of waves and particles holds only in interference experiments (because of the quantitative relationship between intrinsic path distinguishability and mutual coherence in quantum mechanics), but not generally as in tunneling and birefringence (where mutual coherence of the alternative amplitudes plays no pertinent role). Bohr's extension cannot therefore be given the status of a general *principle*. However, all experiments are quite consistent with (i) the Einstein–de Broglie–Bohm version of wave–particle duality [41] in which the position of the particle which is present all the time is guided by the quantum potential determined by the total wave function, as well as with (ii) the causal model for the behaviour of light in terms of a quantum field theory where the hidden variable representing the field is not the photon position but the coordinates of the field modes [42]. Recently a causal model of massive and massless spin-0 and spin-1 bosons (below the threshold of particle production and annihilation) in which the boson position is used as the hidden variable has been developed [43]. In this interpretation, photons travel as particles along the lines of energy flow which are modified by boundary conditions (as in a double-slit configuration), resulting in a re-distribution corresponding exactly to interference and diffraction patterns, as already demonstrated by Prosser [44].

These considerations show that the debate concerning the precise nature of wave–particle duality (waves *and* particles versus waves *or* particles) is not purely metaphysical – it is open to experimental scrutiny and has important bearings on the measurement problem.

2

Cavity quantum electrodynamics

2.1 Introduction

Ever since 1916/17 when Einstein argued [45] [46] that spontaneous emission must occur if matter and radiation are to achieve thermal equilibrium, physicists have believed that spontaneous emission is an inherent quantum mechanical property of atoms and that excited atoms inevitably radiate. This view, however, overlooks the fact that spontaneous emission is a consequence of the coupling of quantized energy states of atoms with the quantized radiation field, and is a manifestation of quantum noise or of emission 'stimulated' by 'vacuum fluctuations'. An infinity of vacuum states is available to the photon radiated by an excited atom placed in free space, leading to the effective irreversibility of such emissions. If these vacuum states are modified, as for example by placing an excited atom between closely spaced mirrors or in a small cavity (essentially the Casimir effect), spontaneous emission can be greatly inhibited or enhanced or even made reversible. Recent advances in atomic and optical techniques have made it possible to control and manipulate spontaneous emission. A whole new branch of quantum optics called 'cavity QED' has developed since 1987 utilizing these dramatic changes in spontaneous emission rates in cavities to construct new kinds of microscopic masers or *micromasers* that operate with a single atom and a few photons or with photons emitted in pairs in a two-photon transition.

The study of the perturbing effect of the environment on atomic states and their radiative properties is useful for the following reasons:

(a) In all realistic experiments the atomic system under study is placed in a finite volume bounded by metallic or dielectric walls, and therefore all precision experiments (such as the measurement of the electron's $g - 2$, the Rydberg constant, etc.) must take into account the perturbing effects of the boundary walls on the system. These effects can be qualitatively understood purely classically in terms of the corrections to the radiation rate and to the frequency of a Rydberg atom oscillating in a cavity due to the field reflected back by the cavity walls. Fortunately, these effects are controllable and can be made very small, which is why QED can be tested so accurately.

(b) On the other hand, the radiative properties of atoms can be greatly modified by placing them close enough to boundaries. The discovery of the micromaser (a single-mode cavity driven by single atoms) has made it possible to study the basic principles of the atom–radiation interaction.

(c) The advantages of the system are that it can be used to generate cavity radiation that exhibits completely new and highly non-classical features such as well-defined numbers of photons (Fock states) and sub-Poissonian statistics. It can also be used to study various aspects of the quantum measurement problem such as to test the complementarity of the interference of single atoms and their intrinsic path distinguishability, to 'erase' information about the 'path' of these atoms and restore atomic coherence and interference, to prepare laboratory 'cousins' of the famous 'Schrödinger cat' and to monitor microscopic fields made of a very small number of photons down to the vacuum field with little or no disturbance.

We will closely follow the material reviewed by Haroche in his 1990 Les Houches lectures [47] which concentrate on the main principles and applications in the microwave domain. The extension to the optical domain is under active development and will be found in the same volume partly in the lectures by Kimble and partly in the lectures on squeezing and optical noise manipulation by Fabre.

2.2 Spontaneous emission rates

2.2.1 *Perturbative low-Q regime*

We will first study the radiative corrections experienced by an atom weakly coupled to a continuum of cavity modes (moderate to low-Q regime). The corrections are then small and can be computed by second order perturbation methods. Consider an excited state $|a\rangle$ of an atom connected to its lower energy states $|j\rangle$ by non-vanishing matrix elements of the electron momentum operator $\vec{p}$. Spontaneous emission corresponds to the transition $|a; vacuum\rangle \rightarrow |j; 1\,photon\,in\,mode\,\mu,\,frequency\,\omega_\mu\rangle$, and is induced by the following term in the atom-field Hamiltonian H_{at}:

$$H_{\mathrm{I}} = -(q/m)\,\vec{A}\cdot\vec{p}, \qquad (2.1)$$

$$\vec{A} = \sum_\mu A_\mu \left[a_\mu \vec{\check{\alpha}}_\mu(r) + a_\mu^\dagger \vec{\check{\alpha}}_\mu^*(r) \right], \tag{2.2}$$

where a_μ and $a_\mu^\dagger$ are the photon annihilation and creation operators in the mode μ, $\vec{\check{\alpha}}_\mu(r)$ are normalized spatial functions which satisfy the cavity boundary conditions, and $A_\mu = (\hbar/2\epsilon_0\omega_\mu)^{1/2}$ are normalization factors which ensure that the field energy density integrated over the cavity volume is $(n + 1/2)\hbar\omega_\mu$ when the cavity contains n photons.

The decay rate depends on the density $\rho^{(cav)}(\omega_{aj})$ of photon modes at frequency ω_{aj}, i.e. the number of modes per unit frequency interval at $\omega = \omega_{aj}$. Describing the bandwidth Γ_μ of each mode in a simple phenomenological way, one can write

$$\rho^{(cav)}(\omega) = \sum_\mu \delta_{\Gamma_\mu}(\omega - \omega_\mu), \tag{2.3}$$

where δ_{Γ_μ} is the normalized Lorentz function of width Γ_μ,

$$\delta_{\Gamma_\mu}(\omega' - \omega_\mu) = \frac{1}{\pi} \frac{\Gamma_\mu/2}{(\omega' - \omega_\mu)^2 + \Gamma_\mu^2/4}. \tag{2.4}$$

Using the Fermi Golden rule, one obtains the total decay rate from level a to j,

$$\gamma_{a \to j}^{(cav)} = \frac{2\pi}{\hbar^2} \sum_\mu |\langle a, 0| H_I | j, 1_\mu \rangle|^2 \, \delta_{\Gamma_\mu}(\omega_{aj} - \omega). \tag{2.5}$$

Using the expression (2.1) for H_I and the expansion (2.2), one gets

$$\gamma_{a \to j}^{(cav)} = \frac{\pi}{\hbar} \frac{q^2}{\epsilon_0} \frac{1}{m^2} \frac{1}{\omega_{aj}} \sum_\mu |\vec{p}_{aj} \cdot \vec{\check{\alpha}}_\mu(R)|^2 \, \delta_{\Gamma_\mu}(\omega_{aj} - \omega). \tag{2.6}$$

where $\vec{p}_{aj}$ is the matrix element of $\vec{p}$ between the states a and jR is the position of the center-of-mass of the atom, and we have replaced ω_μ in the Lorentz functions (2.4) by ω_{aj} because they are δ-like functions of ω. Since $\vec{p} = -(im/\hbar)[\vec{r}, H_{at}]$, it follows that $q\vec{p}_{aj} = -(im\omega_{aj})\vec{D}_{aj}$ where $\vec{D}_{aj} = q\vec{r}_{aj}$ is the matrix element of the atomic dipole operator. Using this result, we finally obtain the spontaneous emission rate in the cavity,

$$\gamma_{a \to j}^{(cav)} = \frac{\pi}{\hbar \epsilon_0} \omega_{aj} \sum_\mu |\vec{D}_{aj} \cdot \vec{\check{\alpha}}_\mu(R)|^2 \, \delta_{\Gamma_\mu}(\omega_{aj} - \omega). \tag{2.7}$$

When the sum over μ approximates to the continuum, this decay becomes for all practical purposes irreversible. If the cavity boundary is removed to infinity, one recovers the familiar free-space emission rate by replacing the sum over μ by the correspondence

$$\frac{1}{L^3} \sum_\mu \sum_{i=1,2} |\vec{\epsilon}_{\mu,i} \cdot \vec{e}|^2 \mapsto \frac{1}{3\pi^2 c^3} \int_o^{+\infty} \omega_\mu^2 \, d\omega_\mu. \tag{2.8}$$

One then obtains

$$\gamma_{a \to j}^{(o)} = \omega_{aj}^3 \, |\, D_{aj}\,|^2 / 3 \, \pi \, \epsilon_0 \, \hbar \, c^3. \tag{2.9}$$

The rate of energy radiation by the atom is obtained by multiplying the rate (2.9) by $\hbar \omega_{aj}$:

$$P_{a \to j}^{(\text{cav})} = \frac{\pi \, \omega_{aj}^2}{\epsilon_0} \sum_\mu |\, \vec{D}_{aj} \cdot \vec{\alpha}_\mu(R)\,|^2 \, \delta_{\Gamma_\mu}(\omega_{aj} - \omega). \tag{2.10}$$

The power dissipation of the classical dipole is identical to this expression if one replaces $2|\, \vec{D}_{aj} \cdot \vec{\alpha}_\mu(R)\,|^2$ by $|\vec{D}_{\text{classical}}(t) \cdot \vec{\alpha}_\mu(r)|^2$. Since the mode dependence is the same in the two expressions, the ratio $\gamma_{a \to j}^{(\text{cav})}/\gamma_{a \to j}^{(o)}$ turns out to be precisely the same as in the case of a classical dipole for any cavity geometry. It is straightforward to calculate this ratio for a classical dipole inside a gap of width L between two plane-parallel perfect conductors by using the method of images [47]. For a dipole placed in the middle of the gap, the infinite sets of image dipoles (one for the 'parallel' ($\parallel$) and one for the 'normal' ($\perp$) configurations) resulting from successive reflections in the conductors, are shown in Fig. 2.1. Figure 2.2 shows the decay rates (normalized to the free-space rates) as a function of the distance between the conductors (mirrors) measured in units of $\lambda/2$. The $\parallel$ and $\perp$ configurations correspond to the solid and dashed lines respectively. The most remarkable feature of the $\parallel$ configuration is the complete suppression of the radiation for $L < \lambda/2$. This is because such radiation cannot form standing waves between the mirrors. For $L = \lambda/2$ the radiation suddenly turns on at three times the free-space value. Similar resonances occur when $L = (2n + 1)\lambda/2$. The decay rate falls to the free-space value as L becomes much larger than λ. In the $\perp$ configuration $\gamma^{(\text{cav})}$ diverges as $L \to 0$. All the dipole images then coincide at one point in phase, resulting in a strong enhancement of the radiation rate. The radiation is then polarized normal to the mirrors along modes which can propagate for arbitrarily small values of L. As L increases, $\gamma^{(\text{cav})}$ oscillates and eventually tends to the free-space value $\gamma^{(o)}$.

Since the density of radiation modes in a cavity of a given geometry is the same for the classical and quantum cases, the relative decay rates $\gamma^{(\text{cav})}/\gamma^{(o)}$ are correctly described by the classical radiation reaction picture. Now, in quantum theory spontaneous emission in free space arises partly from vacuum fluctuations and partly from self-reaction effects [48], [49], [50], [51]. The classical picture of an oscillating dipole corresponds entirely to the self-reaction part. This can be seen if one replaces $|\vec{D}_{\text{classical}}(t)|^2$ by $\langle a\,|D^2\,|\,a\rangle = \sum_j \langle a|\,D\,|j\rangle\langle j\,|D\,|a\rangle$ and treats D_{aj}^2 as the mean-square amplitude of a classical dipole oscillating at frequency $|\omega_{aj}|$. If the state a is more bound than the state j, ω_{aj} is negative, the self-reaction 'decay'

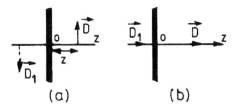

(a)   (b)

Fig. 2.1. A dipole D radiating in front of a mirror. In the parallel configuration (a), the image D_1 is in phase opposition with the dipole. In the normal configuration (b), D_1 is in phase with D (after Ref. [47]).

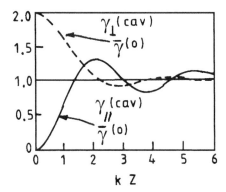

Fig. 2.2. Decay rate of a dipole (normalized to free-space rate) as a function of its distance from a perfectly conducting mirror. Parallel and normal configurations correspond to solid and dashed lines respectively (after Ref. [47]).

is compensated by the vacuum fluctuation 'absorption' and there is no transition from a to j. This provides an explanation for the radiative stability of the atomic ground state in quantum theory. On the other hand, if the state a is more excited than the state j, ω_{aj} is positive and the self-reaction term is only half the exact quantum decay rate; the other half comes from vacuum fluctuations. Since both self-reaction and vacuum fluctuation effects depend in the same way on the cavity geometry, the stability of atomic ground states is not affected by the presence of the walls.

Apart from modifying spontaneous decay rates, the walls of a cavity also cause shifts in the atomic energy levels. For details we refer the reader to Haroche [47].

The experimental study of atomic properties near boundaries started with the pioneering works of Drexhage on dye molecules deposited on a metal surface with a calibrated dielectric spacing between the molecules and the boundary [52]. The experiments clearly demonstrated the change in the fluorescence pattern and decay rates of the molecules. It was

Kleppner [53] who first pointed out the more dramatic changes that are expected when an excited atom is placed in a waveguide structure or in a cavity, such as the complete suppression or enhancement of the spontaneous decay rate. These effects and others associated with them have since been amply demonstrated by a number of experiments using parallel plate cavities, micron-sized gaps, Penning traps, Fabry-Pérot resonators, etc. However, there have been relatively few experiments dealing with atomic energy level shifts. A concise review of these and related experiments with references is given by Haroche [47].

2.2.2 Nonperturbative high-Q regime

Suppose a cavity mode of high quality factor Q and angular frequency ω is resonant or quasi-resonant with a transition of angular frequency ω_0 between two atomic states $|e\rangle$ and $|g\rangle$. The decay rate of the mode, or equivalently its spectral bandwidth is then $\kappa = \omega/Q$. Let the vacuum Rabi frequency of the atomic dipole in the cavity field be

$$\Omega = \frac{|\vec{d}_{eg} \cdot \vec{\mathscr{E}}(r)|}{\hbar}, \tag{2.11}$$

where $\vec{d}_{eg}$ is the dipole matrix element between the atomic levels $|e\rangle$ and $|g\rangle$ and $\vec{\mathscr{E}}(r)$ is the rms vacuum field in the mode at the position r of the atom. Because of the high quality factor Q of the mode, $\Omega > \kappa$ and the perturbative approach (which is valid only when $\Omega \ll \kappa$) breaks down. It is then no longer legitimate to treat the cavity mode as weakly perturbing the atom. Rather it is more appropriate to treat the cavity mode and the relevant atomic transition as a combined 'dressed atom–cavity mode'. The Hamiltonian can be written as

$$H = H_{\text{at}} + H_{\text{F}} + H_{\text{AF}} \tag{2.12}$$

with

$$H\text{at} = \hbar\omega_0 D_3, \tag{2.13}$$

$$H_{\text{F}} = \hbar\omega \left[a^\dagger a + \frac{1}{2} \right], \tag{2.14}$$

$$H_{\text{AF}} = -\hbar\Omega \left[a D^+ + a^\dagger D^- \right], \tag{2.15}$$

where a and $a^\dagger$ are the photon annihilation and creation operators in the cavity mode and D^+, D^- and D_3 are atomic operators:

$$
\begin{aligned}
D^+ &= |e\rangle\langle g|, \\
D^- &= D^{+\dagger} = |g\rangle\langle e|, \\
D_3 &= \tfrac{1}{2}\left[|e\rangle\langle e| - |g\rangle\langle g| \right].
\end{aligned}
\tag{2.16}
$$

The equations (2.12) to (2.16) define the well-known Jaynes–Cummings model in quantum optics which is usually used to describe an atomic system resonantly interacting with a single-mode quantized field containing a very large number of photons. In cavity QED, on the other hand, one considers the more unusual situation in which the cavity contains no photons (the vacuum) or a small number of them. The energy levels of a two-level atom and those of the quantized cavity mode (harmonic oscillator levels) are separately shown in Fig. 2.3. Figure 2.4 shows the energy levels of the 'dressed atom–cavity mode' system. The ground state is $|g, 0\rangle$, representing the atom in its ground state and the cavity mode in its vacuum state. The excited energy eigenstates form an infinite set of two-state manifolds separated by $\hbar\omega$, each pair split by an amount proportional to the vacuum Rabi frequency Ω and to the square root of the photon number of the corresponding cavity mode. The dressed energy eigenstates in the nth manifold are linear combinations of the states $|e, n\rangle$ and $|g, n+1\rangle$ [47]:

$$|\overline{+, n}\rangle = \cos\theta_n|e, n\rangle + \sin\theta_n|g, n+1\rangle, \tag{2.17}$$

$$|\overline{-, n}\rangle = -\sin\theta_n|e, n\rangle + \cos\theta_n|g, n+1\rangle, \tag{2.18}$$

where the coupling angle θ_n is defined by

$$\tan 2\theta_n = \frac{2\Omega\sqrt{n+1}}{\omega_0 - \omega} \quad (0 \le \theta_n < \pi/2). \tag{2.19}$$

The energies of the dressed states are

$$E_{g0} = -\frac{\hbar}{2}(\omega_0 - \omega), \tag{2.20}$$

$$E_{\overline{\pm, n}} = (n+1)\hbar\omega \pm \frac{\hbar}{2}\Delta_n, \tag{2.21}$$

with

$$\Delta_n = \sqrt{4\Omega^2(n+1) + (\omega_0 - \omega)^2}. \tag{2.22}$$

At exact resonance ($\omega_0 = \omega$ and $\theta_n = \pi/4$), one has

$$|\overline{+, n}\rangle = \frac{1}{2}[|e, n\rangle + |g, n+1\rangle], \tag{2.23}$$

$$|\overline{-, n}\rangle = \frac{1}{2}[-|e, n\rangle + |g, n+1\rangle], \tag{2.24}$$

and

$$\Delta_n(\omega_0 = \omega) = 2\Omega\sqrt{n+1}. \tag{2.25}$$

Suppose at $t = 0$ the atom–field system is prepared in the state $|e, 0\rangle$ which is a linear combination of the dressed energy eigenstates $|\overline{+, 0}\rangle$ and

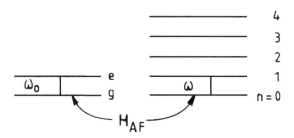

Fig. 2.3. Energy levels of a two-level atom and a quantized cavity mode. The two subsystem frequencies ω and ω_0 are equal, or very close (after Ref. [47]).

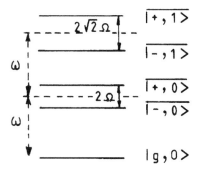

Fig. 2.4. Energy diagram of the 'dressed atom–cavity mode' system (after Ref. [47]).

$|-,0\rangle$. In the absence of relaxation, these energy eigenstates evolve in time according to [47]

$$|\Psi(t)\rangle = \cos\theta_0\, e^{-iE_{\overline{+,0}}\, t/\hbar}|\overline{+,0}\rangle - \sin\theta_0\, e^{-iE_{\overline{-,0}}\, t/\hbar}|\overline{-,0}\rangle \qquad (2.26)$$

which exhibits oscillations. Therefore the probability $P_e(t)$ of finding the atom in level e at time t is obtained by projecting $|\Psi(t)\rangle$ onto $|e,0\rangle$:

$$P_e(t) = \cos^2 \Omega t. \qquad (2.27)$$

These oscillations are damped or overdamped when cavity relaxation is taken into account. When cavity relaxation is strong (low-Q regime), one recovers the perturbative result (exponential decay of $P_e(t)$).

The nonperturbative regime describes a *reversible* energy exchange between the atom and the cavity mode, quite unlike the irreversible spontaneous photon emission process described in the previous section. Like the irreversible spontaneous emission case, the reversible vacuum Rabi oscillations can also be understood *qualitatively* in terms of classical electrodynamics [47], though a full understanding of them requires the incorporation of vacuum fluctuations which are of purely quantum theoretical origin.

The Rabi oscillations in the vacuum and in a weak coherent field have been observed [54]. Ever since Planck proposed his law of black-body radiation and Einstein offered his simple explanation of the features of photoelectricity in terms of light quanta, quantization of the radiation field has been widely accepted. As we saw in the previous Chapter, the photoelectric effect, although generally regarded as evidence of field quantization, can also be explained in terms of a classical field description, provided the linear detector is a quantum system. Many other phenomena such as the Compton effect, spontaneous emission and radiative QED corrections have also pointed to the existence of field quanta. The demonstration of non-classical field behaviours such as anti-bunching [55], [56], sub-Poissonian photon statistics [57] and squeezing [58], [59] in quantum optics is also strongly suggestive of field quantization. However, none of these observations could be claimed as direct evidence of field quanta. As we saw in the previous Chapter, the production of single-photon states of light in the laboratory and the observation of anti-coincidence on a beam splitter provided the first direct evidence of field quantization. The importance of the observation of Rabi oscillations lies in providing the first ever direct evidence of energy quantization in a cavity mode. To understand the principle of the experiment, we need first to understand how the micromaser works.

2.3 Micromasers

Figure 2.5 is a sketch of a Rydberg atom micromaser setup. A highly collimated beam of atoms is prepared in the excited Rydberg level e by selective laser excitation and made to pass *one at a time* through a superconducting cavity with a high quality factor and tuned to resonance with the atomic transition from e to a neighbouring lower energy state g. The cavity is liquid helium cooled to suppress black-body induced transitions. After emerging from the cavity the atoms are detected by two separate electric field zones which can discriminate between the states e and g. There are three different time scales involved in the operation of the micromaser. One is the atom–cavity interaction time t_{int} which can be controlled by first selecting the velocities of the atoms before they are laser excited by means of a set of spinning wheels with slots which intercept the atomic beam (a Fizeau velocity selector). For a thermal beam with a mean velocity $\bar{v}$ the mean value of this interaction time is $\bar{t}_{int} = L/\bar{v}$, where L is the length of the cavity. It is possible to select the velocities such that the micromaser can be operated with fixed, well-defined and tunable interaction times t_{int}. Typical values for cm-size cavities and thermal velocities (100 to 300 m/s) are 20 to 100 μs. Another scale is the mean

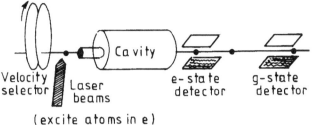

Fig. 2.5.   Sketch of a Rydberg atom micromaser setup (after Ref. [47]).

time t_{at} between consecutive atoms in state e in the cavity, which can be controlled by adjusting the pump laser power. By varying the laser power from zero to the maximum, t_{at} can be varied between infinity and 10^{-7} to 10^{-8} s. Finally, there is the cavity photon damping time $t_{cav} = \kappa^-$ where κ is the cavity decay rate ω/Q defined earlier. Typical values for $Q \approx 10^{10}$, $\omega \approx 20$–70 GHz and temperatures in the range of a Kelvin or less are $t_{cav} = 0.2$ to 10^{-3} s.

It is convenient to define the average number of atoms crossing the cavity during its decay time by $N_{ex} = t_{cav}/t_{at}$. For true micromaser operation one must ensure that there is no more than one atom at a time in the cavity. This is guaranteed if $t_{at} > t_{int}$. This condition can be satisfied with $N_{ex} \gg 1$ if

$$t_{int} < t_{at} < t_{cav}. \tag{2.28}$$

The values of the parameters given above satisfy this condition.

The vacuum Rabi frequency $\bar{\Omega}$ averaged over the atom path across the cavity defines the coupling of the atoms with the cavity mode. For low-order mode cavities and Rydberg numbers around 40 to 65, $\bar{\Omega} \approx 10^4 - 10^5$ s^{-1}. At resonance the probability for an atom of velocity v to undergo the transition $e \to g$ in a cavity containing n photons is given by

$$\pi(n) = \sin^2\left(\bar{\Omega}\sqrt{n+1}\, t_{int}\right). \tag{2.29}$$

Therefore, when the first excited atom in the Rydberg state e enters the cavity which is initially empty ($n = 0$), it decays to the state g with probability $\pi(0) = \sin^2(\bar{\Omega} t_{int})$, releasing a photon into the cavity. If the micromaser operates in an intrinsically nonlinear regime, i.e. if $\bar{\Omega} t_{int} \geq 1$, this probability is quite large (of the order of unity). The second excited atom must enter the cavity before the first photon decays. This is guaranteed since $t_{at} < t_{cav}$ (2.28). The second atom will interact with the field and emit a second photon of the same frequency with probability $\pi(1) = \sin^2(\sqrt{2}\bar{\Omega} t_{int})$ of order unity. This process must continue for the subsequent atoms. The steady state field in the cavity eventually results from the competition between gain from photon emissions by the atoms with probability $\pi(n)$ of order unity and cavity losses. Since the field at any time t is produced by the atoms that have crossed the cavity during an interval of time of the order of t_{cav} prior to t, the average number $\bar{n}$ of photons in the cavity is of the order of N_{ex}. With the orders of magnitude of the parameters discussed above, N_{ex} and $\bar{n}$ are $\approx 10^2 - 10^3$. This shows the remarkable character of a micromaser: it operates with only one atom in the cavity at a time and a few tens or hundreds of photons.

The liquid helium cryostat is carefully designed to prevent room temperature microwave photons from leaking into the cavity, which would considerably increase the temperature of the cavity radiation above the temperature of the cavity walls. The cavity is also carefully shielded against magnetic fields by several layers of cryoperm™. In order to achieve the high quality factor required and prevent the different magnetic substates of the maser levels from mixing during the atom–field interaction time, three pairs of Helmholtz coils are used to compensate the earth's magnetic field to a value of several mG in a volume of $10 \times 4 \times 4$ cm^3.

It is also possible [47] to operate micromasers based on two-photon transitions between levels e and g having the same parity, the two-photon transition occurring via an intermediate virtual transition to a level i having opposite parity.

Mostly, methods of nonlinear optics are used to generate non-classical radiation. The micromaser is another source. By controlling the interaction time t_{int} it is possible to realize micromasers in which the field statistics is strongly sub-Poissonian or super-Poissonian. It is, of course, not possible to measure this field directly without affecting the high-Q character of the cavity and consequently the non-classical properties of the field itself. However, information about the field can be gleaned by detecting the quantum state of the atoms emerging from the cavity. As shown in Fig. 2.5, the atoms successively traverse two detecting zones in which they are differentially ionized. In the first zone the electric field is strong enough to ionize only the upper level e but not the more strongly bound level g. In the second zone, on the other hand, the field is strong enough to ionize the level g. In this way it is possible to detect the numbers N_e and N_g of atoms in the levels e and g respectively that emerge from the cavity in a given time. The statistical properties of these numbers exhibit interesting non-classical features, reflecting the non-classical statistics of the radiation in the cavity. A sub-Poisssonian variance of N_g means that the arrival times of atoms in level g are anti-bunched, i.e. the conditional probability of detecting an atom in level g immediately after one has been detected in this level is smaller than the average unconditional probability. In contrast, a bunching of atoms in the detector signals a super-Poissonian variance. For further details of the statistics of the radiation field in the cavity and of the emerging atoms, see Haroche [47] and the references therein.

The fact that the same atoms act as the source and detector of the cavity field is a feature that introduces complications in the measurement process. The atoms interact with the field in a resonant way and change the very state of the field one wishes to measure. This is the well-known and inevitable quantum back-action of the measuring device (the atoms) on the system to be measured (the field). Using a low flux of atoms is of no help when the number of photons in the cavity is extremely small (of the order of unity). There is, however, a way of avoiding this back-action that enables one to monitor the cavity field without changing its Fock state or photon number – a quantum nondemolition (QND) measurement, as we will see in the next Chapter. The combination of QND and cavity QED methods has opened up the possibility of doing novel types of experiments in which the state of a microscopic field made of a very small number of photons down to the vacuum field can be monitored with minimal disturbance. For details, see, for example, the paper by Brune *et al.* [60].

The setup to observe the Rabi oscillations is shown in Fig. 2.6. Rubidium atoms effusing from the oven O are prepared in the circular Rydberg state *e* with principal quantum number 51 in the box B. At a repetition rate of 660 Hz, 2 μs long pulses of Rydberg atoms start from B with a Maxwellian velocity distribution (mean velocity $v_0 = 350$ m/s). The atoms cross the cavity C made of niobium superconducting mirrors and are detected by state-selective field ionization by the detector D, enabling the transfer rate from *e* to *g* (principal quantum number 50) to be measured. The cavity is tuned into resonance with the *e* to *g* transition (frequency 51.099 GHz). The cavity *Q* factor is 7×10^7, corresponding to a photon life-time $T_{cav} = 220$ μs, which is longer than the atom–cavity interaction time. In order to avoid field buildup by cumulative atomic emission, the average delay between successive atoms was adjusted to be 2.5 ms, much longer than T_{cav}. Each atom therefore felt a field restored by S (a very stable source used to inject continuously into the cavity a small coherent field with a controlled energy varying from zero to a few photons) to its initial state. Taking into account the detection efficiency, the actual counting rate was 30 s^{-1}. The atom–cavity interaction time was carefully controlled by determining the velocity of each detected atom to an accuracy of 1% from a knowledge of its arrival time in D and of its preparation time in B. Atoms of a range of velocities were selected and the signals $P_{e,g}(t)$ plotted as shown in Fig. 2.7 (A)–(D). The signals in Fig. 2.7(A) exhibit the reversible spontaneous emission and absorption of a single photon in an initially empty cavity mode, an effect predicted by the Jaynes–Cummings model but never observed before in the time domain. Figure 2.7(B)–(D)

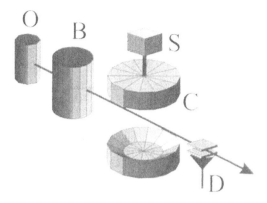

Fig. 2.6. Sketch of the experimental setup to observe Rabi oscillations (after Ref. [54]).

shows that when a small coherent field is injected, the signal is no longer sinusoidal, as it would be for an atom in a classical field. After a first oscillation, a clear collapse and revival feature is observed in Figure 2.7(C) and (D). The oscillation dampings are due mainly to dark counts in the ionization detectors.

Figure 2.7(a)–(d) shows the Fourier transforms of the nutation signals, obtained after symmetrization with respect to $t = 0$. Discrete peaks at frequencies $v = 47$ kHz, $v\sqrt{2}$, $v\sqrt{3}$ and even $2v$ are clearly observable, revealing directly the quantized nature of the field up to three photons. The frequency v is in good agreement with the expected value 50 kHz.

The solid curves in Fig. 2.7 (A)–(D) were obtained by fitting the time-dependent signals with a sum of damped sinusoids with frequencies $v\sqrt{n+1}$, n varying from 0 to 5. The photon number probabilities were determined from the relative weights of the terms in these fits, and are shown as Fig. 2.7(α)–(β). When no field is injected [Fig. 2.7(α)], there is very good agreement with the thermal radiation law (solid line) with the very small average photon number $\bar{n} = 0.06(\pm0.01)$, corresponding well with the average value deduced from the cavity temperature (0.05 photon at $T = 0.8$ K). When a small coherent field is injected [Figs. 2.7(β)–(δ)], there is very good agreement between the experimental data and a Poissonian law (solid lines), providing an accurate value of the mean photon number in each case: $0.40(\pm 0.02)$, $0.85(\pm 0.04)$ and $1.77(\pm 0.15)$ respectively.

The existence of the several components of the spectrum imply the increase of the atom–field coupling in discrete steps, and is a quantum nonlinear effect. Such effects were observed in this experiment with less than half a photon on average. It is this nonlinearity of the atom–

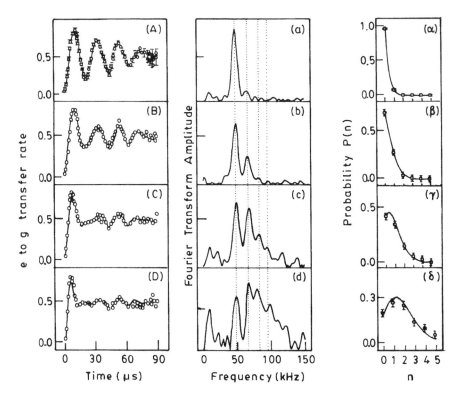

Fig. 2.7. (A)–(D): Rabi nutation signals for fields with increasing amplitudes. (A): No injected field, 0.06(± 0.01) thermal photons on average; (B)–(D): coherent fields with 0.40(± 0.02), 0.85(± 0.04) and 1.77(± 0.15) photons on average. The points are experimental and the solid curves are theoretical fits. (a)–(d): Corresponding Fourier transforms. (α)–(β): Corresponding photon number distributions. The solid curves show the theoretical thermal (α) or coherent [(β)–(δ)] distributions which best fit the data (after Ref. [54]).

field coupling at very low field strengths that makes the Rydberg atom sensitive to the quantum behaviour of the field, and distinguishes it from an ordinary linear detector which is not sensitive. The combination of this type of resonant method with dispersive field detection methods [61] using Rydberg atoms and microwave cavities has opened up the possibility of many fascinating applications to measurement and manipulation of weak quantum fields and information processing [62].

2.3.1 *Mesoscopic Schrödinger cats*

The features of a micromaser discussed above make it possible to use it to prepare linear superpositions of two different coherent states (minimum uncertainty states that are closest to classical states) that are highly non-classical in nature. They are laboratory 'cousins' of the famous cat that Schrödinger conceived of to highlight the paradoxes that arise when 'an indeterminacy originally restricted to the atomic domain becomes transformed into macroscopic indeterminacy, which can then be *resolved* by direct observation' [63]. Here is how Schrödinger introduced the paradox:

> A cat is penned up in a steel chamber, along with the following diabolical device (which must be secured against direct interference by the cat): in a Geiger counter there is a tiny bit of radioactive substance, *so* small, that *perhaps* in the course of one hour one of the atoms decays, but also, with equal probability, perhaps none; if it happens, the counter tube discharges and through a relay releases a hammer which shatters a small flask of hydrocyanic acid. If one has left this entire system to itself for an hour, one would say that the cat still lives *if* meanwhile no atom has decayed. The first atomic decay would have poisoned it. The ψ-function of the entire system would express this by having in it the living and the dead cat (pardon the expression) mixed or smeared out in equal parts.

Various aspects of cat states have been reviewed by Glauber [64]. Although it is perhaps impossible to prepare a real cat in a linear superposition of 'alive' and 'dead' states, it is certainly possible to produce their laboratory 'cousins' by using various nonlinear processes [65] such as propagation through a Kerr medium [66], the use of micromasers [67] and quantum nondemolition methods [68]. They are expected to show non-classical features [69]. Micromasers are specially suited to produce such superpositions using coherent states of radiation

$$|\alpha\rangle = \sum_n c_n(\alpha)|n\rangle \tag{2.30}$$

with

$$c_n(\alpha) = \frac{\alpha^n}{\sqrt{n!}} e^{-|\alpha|^2/2}, \tag{2.31}$$

and thus allow the exploration of the fuzzy boundary between the classical and quantum domains. Suppose the initial field in the cavity is described by such a state. Let each atom be prepared by laser and microwave irradiation in the circular Rydberg level e before it enters the cavity. It then interacts in a zone R_1 with an auxiliary microwave field tuned at a frequency ω_r resonant or quasi-resonant with the $e \rightarrow g$ transition. The atom then leaves this zone in a linear superposition of the states e and g.

For an atom that has undergone a $\pi/2$ pulse, this state is

$$|\Psi_i\rangle = \frac{1}{\sqrt{2}} \left[|e\rangle + |g\rangle \right]. \tag{2.32}$$

Let such a state then traverse the cavity, excite it and then interact with a second auxiliary microwave field, identical to the first one, in a second region R_2. If the atom has velocity v, the state of the total atom–field system before detection of the atom is then described by

$$|\Psi_T\rangle = \sum_n c_n(\alpha) \left[b_e(v, n)|e, n\rangle + b_g(v, n)|g, n\rangle \right] \tag{2.33}$$

where b_e and b_g are appropriate probability amplitudes. If the auxiliary fields (called Ramsey fields) are exactly resonant and the atom has a velocity v_0 and undergoes a $\pi/2$ pulse in each zone, one can show [47] that (2.32) takes the form

$$|\Psi_T\rangle = \frac{1}{2} \sum_n \left[c_n(\alpha) [e^{i\phi_n} - 1] |e, n\rangle - c_n(\alpha) [e^{i\phi_n} + 1] |g, n\rangle \right] \tag{2.34}$$

where $\phi_n = n\epsilon$ is the phase difference between the atomic dipole and the auxiliary Ramsey field reference in the presence of n photons in the cavity. If the atom is detected in the state e, this state reduces (within a normalization factor) to

$$|\Psi(e)\rangle \simeq \sum_n c_n(\alpha) [e^{-in\epsilon} |n\rangle - |n\rangle] \tag{2.35}$$

$$\simeq |\alpha e^{-i\epsilon}\rangle - |\alpha\rangle \tag{2.36}$$

because $c_n(\alpha) e^{-in\epsilon} = c_n(\alpha e^{-i\epsilon})$. Notice that this is a linear superposition of the initial state and the state obtained by shifting its phase by ϵ. If, on the other hand, the atom is detected in the state g, the state (2.34) collapses to the linear superposition

$$|\Psi(g)\rangle \simeq |\alpha e^{-i\epsilon}\rangle + |\alpha\rangle. \tag{2.37}$$

In the particular case when $\epsilon = \pi$, (2.36) and (2.37) take the form

$$|\Psi(e)\rangle = \frac{1}{2 \sum_{n\,\mathrm{odd}} |c_n|^2} [|-\alpha\rangle - |\alpha\rangle] \tag{2.38}$$

and

$$|\Psi(g)\rangle = \frac{1}{2 \sum_{n\,\mathrm{even}} |c_n|^2} [|-\alpha\rangle + |\alpha\rangle]. \tag{2.39}$$

What has transpired is the following: the linear superposition of states e and g (2.32) carried by the atom has been transferred to the cavity field after the atom interacts with the second Ramsey zone and is detected. The situation is similar to the Schrödinger cat paradox in which the linear

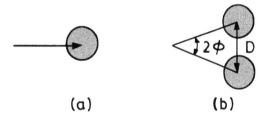

(a)   (b)

Fig. 2.8. (a) Pictorial representation in phase space of a coherent state of a quantum oscillator. (b) The two components separated by a distance D of a Schrödinger cat (after Ref. [70]).

superposition of the radioactive atom (in the decayed and undecayed states) is transferred to the macroscopic cat in the chamber. If all the atoms have the same velocity v_0 and $\epsilon = \pi$, all of them leave the cavity in the same state and the cat state in the cavity is stable. If the atomic velocities are different, each component of the cat state will experience different phases, leading eventually to a phase diffusion process and a totally random phase distribution.

Schrödinger cat states have actually been realized in the laboratory using these principles [70]. If the $e \rightarrow g$ transition and the cavity frequency are slightly off-resonance (detuning δ), the atom and the field cannot exchange energy but only undergo $1/\delta$ dispersive frequency shifts. The c-number α which defines the coherent state of the cavity oscillator can be regarded as the 'meter' or the 'cat', and can be represented by a vector in phase space ($\alpha = \sqrt{n}$ where n is the mean number of oscillator quanta). Quantum fluctuations make the tip of this vector uncertain, with a circular gaussian distribution of radius unity (Fig. 2.8). The 'atom–meter' coupling during time t produces an atomic-level dependent dephasing of the field and entangles the phase of the oscillator ($\pm \phi$) to the state of the atom, leading to the combined state

$$|\Psi\rangle = \frac{1}{\sqrt{2}} \left(|e, \alpha e^{i\phi}\rangle + |g, \alpha e^{-i\phi}\rangle \right), \tag{2.40}$$

with $\phi = \Omega^2 t/\delta$ for $\Omega/\delta \ll 1$. When the 'distance' $D = 2\sqrt{n}\sin\phi$ between the meter states is larger than 1, a Schrödinger cat state is obtained (Fig. 2.8(b)).

In the actual experiment (Fig. 2.9) the states e and g were circular Rydberg levels of rubidium atoms with principal quantum numbers 51 and 50, corresponding to the transition frequency $v_0 = 51.099$ GHz. They have a very long radiative life-time (30 ms) and couple strongly to radiation. The cavity C was a Fabry–Pérot resonator with superconducting niobium mirrors, a Q factor of 5.1×10^7 (photon life-time $T_r = 160$ µs)

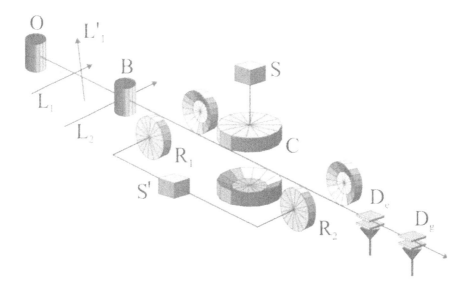

Fig. 2.9. Sketch of the experimental setup (after Ref. [70]).

and $\Omega/2\pi = 24$ kHz. It contained a small coherent field with an average photon number n varying between 0 and 10 injected by a pulsed source S. The axis of the cavity was kept normal to the atomic trajectory, and it was tuned by adjusting the mirror separation so that $\delta/2\pi$ could be varied between 70 and 800 kHz. The effective interaction time was set to 19 µs by selecting atoms with a velocity of 400 m/s.

R_1 and R_2 were low-Q cavities fed by a cw source S' whose frequency ν could be swept across ν_0. The atoms were finally detected and counted by two field ionization detectors D_e and D_g with detection efficiency $40 \pm 15\%$. With 50 000 events recorded in 10 minutes, the probability $P_g^{(1)}(\nu)$ to find the atom in the state g as a function of ν could be reconstructed.

Since the $e \rightarrow g$ transition can occur either in R_1 (atom crossing C in state g) or in R_2 (atom crossing C in state e) and these two 'paths' are indistinguishable, the corresponding amplitudes interfere. The phase difference between these amplitudes is $2\pi(\nu - \nu_0)T$ where $T = 230$ µs is the time interval between the $\pi/2$ pulses to which the atoms are subjected in R_1 and R_2. Thus, $P_g^{(1)}(\nu)$ is expected to oscillate with the frequency $1/T = 4.2$ kHz. Figure 2.10(a) shows the signal $P_g^{(1)}(\nu)$ exhibiting these Ramsey fringes with C empty and $\delta/2\pi = 712$ kHz. Figure 2.10(b)– (d) correspond to a coherent field in C with $|\alpha| = \sqrt{9.5} = 3.1$ and $\delta/2\pi = 712, 347$ and 104 kHz respectively.

When an atom leaves C, the system is prepared in the entangled state (2.40) and *the field phase 'points' towards e and g at the same time*. The

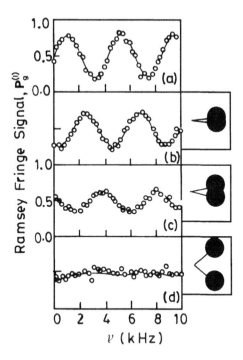

Fig. 2.10. $P_g^{(1)}(v)$ signal exhibiting Ramsey fringes: (a) C empty, $\delta/2\pi = 712$ kHz; (b)–(d) C stores a coherent field with $|\alpha| = 3.1$, $\delta/2\pi = 712$, 347 and 104 kHz respectively. Points are experimental and curves are sinusoidal fits. Insets show the phase space representation of the field components left in C (after Ref. [70]).

insets in Fig. 2.10 show the phase space representation of the two field components. When δ is large (i.e. ϕ small), the measurement of the field phase gives no information of the state of the atom in C because of the large overlap of the field components – the atom crosses C in e or g and these 'paths' being still indistinguishable, the contrast remains large (Fig. 2.10(b)). As δ is reduced (ϕ increased), the field contains more and more information of the state of the atom in C, the two 'paths' become more and more distinguishable and the fringe contrast decreases (Fig. 2.10(c)). It vanishes when the field components do not overlap at all (Fig. 2.10(d)).

Note that the mere fact that the atom leaves in C an information that *could* be read out destroys the interference – it is not actually necessary to measure the field. This once again illustrates the basic aspect of complementarity discussed in the previous Chapter. The fringe contrast is expected to be reduced by a factor equal to the modulus of the overlap integral $\langle \alpha e^{-i\phi} | \alpha e^{i\phi} \rangle = \exp(-D^2/2)\exp(i n \sin 2\phi)$. This is in very

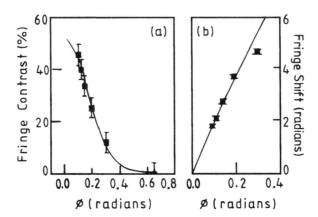

Fig. 2.11.   (a) Fringe contrast and (b) fringe shift versus ϕ for a coherent field with $|\alpha| = 3.1$ (points: experiment; line: theory) (after Ref. [70]).

good agreement with the measured points (Fig. 2.11(a)). Further, a simple consideration shows that the phase of the fringes is shifted by an angle equal to the phase of the same overlap integral, namely $n \sin 2\phi$. This shows that the fringe shift is proportional to n which can therefore be determined from this set of data. Figure 2.11(b) shows that the best value is $n = 9.5 \pm 0.2$.

Quantum coherence between the two components of the state and their progressive decoherence were verified in a subsequent two-atom correlation experiment. The first atom traversing C creates a superposition of states involving two field components. The second 'probe' atom crosses C with the same velocity after a short delay τ and dephases again the field by an angle $\pm \phi$, turning the two field components into three with phases $\pm 2\phi$ and zero. The zero-phase component is produced because the atoms may have crossed C either in the (e, g) or in the (g, e) configurations. Since the atomic states are mixed after C in R_2, these two paths are indistinguishable. Consequently, the joint probabilities $P_{ee}^{(2)}, P_{eg}^{(2)}, P_{ge}^{(2)}$ and $P_{gg}^{(2)}$ of detecting any of the four possible two-atom configurations interfere. It is possible to calculate these probabilities with a few simplifying assumptions. It turns out that the parameter

$$\eta = \left[P_{ee}^{(2)} / (P_{ee}^{(2)} + P_{eg}^{(2)}) \right] - \left[P_{ge}^{(2)} / (P_{ge}^{(2)} + P_{gg}^{(2)}) \right] \qquad (2.41)$$

is independent of v except around $\phi = 0$ and $\phi = \pi/2$. At short times when quantum coherence is fully preserved, $\eta = 0.5$. However, according to decoherence models which we will discuss in greater detail in Chapters 5 and 7, the 'meter' coordinate is invariably coupled to a large reservoir of microscopic variables which, when traced over, induce a fast dissipation

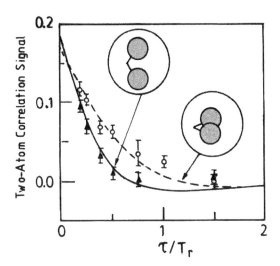

Fig. 2.12. ν-averaged η values versus τ/T_r for $\delta/2\pi = 170$ kHz (circles) and $\delta/2\pi = 70$ kHz (triangles). Dashed and solid lines are theoretical. Insets: pictorial representations of corresponding field components separated by 2ϕ (after Ref. [70]).

of macroscopic coherence. Therefore, as the 'atom + field' system evolves into an effectively incoherent statistical mixture, η should decay to zero.

The parameter η was measured by replacing the Rydberg state preparation pulse by a pair of pulses separated by τ, varied between 30 and 250 μs. The sequence was repeated every 1.5 ms and statistics on double detection events were accumulated. For each delay τ, 15,000 coincidences were detected. Figure 2.12 shows a plot of $\bar{\eta}$ (η averaged over ν) versus τ/T_r for $n = 3.3$ and two different detunings ($\delta/2\pi = 170$ and 70 kHz). It is clear that the correlation signals decrease with time, revealing the dynamics of quantum decoherence. Also, decoherence was observed to proceed faster when the 'distance' between the two state components was increased (as indicated by the insets in Fig. 2.12), in agreement with predictions of simple decoherence models. The significance of these observations will become clearer later from the discussions in Chapters 5 and 7. We will then be in a better position to appreciate to what extent this experiment provides 'a direct insight into a process at the heart of quantum measurement'.

A mesoscopic Schrödinger cat state of a material oscillator has also been generated by preparing a laser-cooled trapped single ^{9}Be$^+$ ion in a superposition of two spatially separated coherent harmonic oscillator states (motional states of the atom) entangled with its internal electronic quantum states [71]. This cat-like superposition was verified by detection of the quantum mechanical interference between the localized wave packets, but no decoherence experiment was carried out in this case.

2.3.2 Testing complementarity

One of the aims of cavity QED is to prepare Fock states of a microwave field in a high-Q cavity. Let a monokinetic beam of atoms prepared in the Rydberg state e successively cross two identical cavities C_1 and C_2 tuned to resonance with an atomic transition between e and a nearby lower energy state g. To keep to the essentials we will assume that Q is infinite so that there is no relaxation. To start with the two cavities are prepared such that C_1 is empty ($n = 0$) and C_2 has only one photon ($n = 1$) in it. Then initially the atom–field system is in the state $|e; 0, 1\rangle$. Let the velocity of the atoms be so tuned that the atom receives a π-pulse in C_1 and a reverse π-pulse in C_2. Then it will emit a photon with unit probability into C_1 and then absorb the photon in C_2 with unit probability. The net effect will be that the atom will revert back to its state e after crossing the two cavities, and the photon will be transferred from C_1 to C_2 with unit probability. Thus, although the atomic state remains the same, it leaves behind a 'tag' signalling its passage through the cavities. So the final state of the atom–field system will be $|e; 1, 0\rangle$.

Now suppose that the Rydberg states e and g are both degenerate with sublevels $|e_+\rangle, |e_-\rangle$ and $|g_+\rangle, |g_-\rangle$. These could, for example, be the spin $1/2$ states quantized along the z-direction. Assume that both the cavities sustain a σ^+ circularly polarized mode that can induce the transition $|e_+\rangle \rightarrow |g_-\rangle$ but leaves the substate $|e_-\rangle$ unaffected. Let the atom be initially prepared in the state

$$|\Phi_0\rangle = \frac{1}{\sqrt{2}} \big[|e_+\rangle + |e_-\rangle \big] \tag{2.42}$$

which is a linear superposition with a definite phase relation between the states $|e_+\rangle$ and $|e_-\rangle$. This state is an eigenstate of the operator σ_x with eigenvalue $+1$. Therefore if this state is sent through a Stern–Gerlach magnet aligned along the x direction followed by two detectors D_+ and D_- which can count atoms with their spins in the $+1/2$ and $-1/2$ states relative to the x-axis (Fig. 2.13), the expectation value $\langle \sigma_x \rangle$ being $+1$, only the detector D_+ will register clicks. The probability amplitudes corresponding to the two 'paths' followed by the states $|e_+\rangle$ and $|e_-\rangle$ through the Stern–Gerlach system, interfere constructively in D_+ and destructively in D_- before being detected.

Imagine a different experiment in which the same atomic state $|\Phi_0\rangle$ is first sent through the cavities C_1 and C_2. Then the state of the atom-field

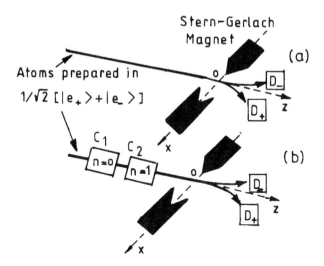

Fig. 2.13. Sketch of a *gedanken* Stern–Gerlach experiment illustrating complementarity. (a) Atoms prepared in a coherent superposition of levels e_+ and e_- are in an eigenstate S_x and are thus all detected by detector D_+. (b) The atoms traverse the C_1–C_2 cavity system containing Fock states $n = 0$ and $n = 1$, which results in the 'tagging' of level e_+; the atomic coherence is destroyed, resulting in equal counts by detectors D_+ and D_- (after Ref. [47]).

system is changed to

$$|\Psi_i\rangle = \frac{1}{\sqrt{2}} \left[|e_+; 0, 1\rangle + |e_-; 0, 1\rangle \right]. \tag{2.43}$$

Since the cavities are resonant with the $|e_+\rangle \to |g_-\rangle$ transition alone, only the $|e_+\rangle$ state is tagged and not the $|e_-\rangle$ state. Consequently, the state $|\Psi_i\rangle$ evolves into

$$|\Psi_f\rangle = \frac{1}{\sqrt{2}} \left[|e_+; 1, 0\rangle + |e_-; 0, 1\rangle \right]. \tag{2.44}$$

The expectation value of σ_x in this state is $\langle \sigma_x \rangle = 0$. This is because the field states $|1, 0\rangle$ and $|0, 1\rangle$ being orthogonal, the density operator of the atomic system, obtained by tracing over the field states, is now diagonal. In other words, the coherence of the atomic states is now destroyed. As a result, both the detectors D_+ and D_- will register clicks. This change in the Stern–Gerlach analyzer pattern results from the fact that the information about the different 'paths' followed by the states $|e_+\rangle$ and $|e_-\rangle$ is stored in the cavities. This provides a test of complementarity between interference and 'which path' information. Notice that the atom itself has not been uncontrollably 'disturbed' by the detection of its path, yet the coherence has been destroyed simply by the information stored in the cavities. Such

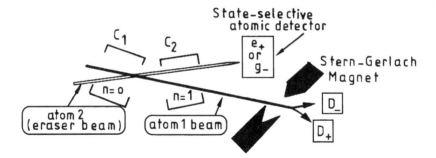

Fig. 2.14. Sketch of a *gedanken* Stern–Gerlach experiment illustrating the principle of the 'quantum eraser' experiment. In the absence of atom 2, detectors D_+ and D_- register equal counts. An atom 2 is sent into the system after each atom 1 and is detected by a state-selective detector. If only the events associated with one type of atom 2 measurement (e_+ or g_-) are selected, coherence in the atom 1 signal is restored and uneven counts are obtained in D_+ and D_- (after Ref. [47]).

a result would not be possible with coherent Glauber states in the cavities C_1 and C_2 because the final states produced by the addition or subtraction of a single photon would not be orthogonal to each other. Consequently, the coherence of the atomic states would not be destroyed and $\langle \sigma_x \rangle$ would not vanish. This underscores the crucial role played by the quantum Fock states in the cavity and their changes in testing complementarity. (For further discussions on complementarity, see Chapter 1.)

2.3.3   *The quantum eraser*

The question that one may ask at this stage is whether the information about the system 'path' stored in the cavities can be 'erased' later at any time before the atoms are finally detected in D_1 and D_2 ('delayed choice') and the coherence and interference of the atomic states restored. The answer is yes, because the path detector is a quantum system and not a classical one. (The concept of a quantum eraser was first proposed by Scully [72], [5]). This can be done by sending a second atomic beam through the cavities to scramble the information left by the first beam. The principle is sketched in Fig. 2.14. Let atom 2 which is identical to atom 1 be introduced into the cavities in the state $|g_-\rangle$. After it has interacted with the two cavities, the total system (atom 1 + atom 2 + field) is in the final state

$$|\Psi_f\rangle = \frac{1}{\sqrt{2}} \left[|e_+\rangle_1 \otimes U(t)|g_-; 1,0\rangle + |e_-\rangle_1 \otimes U(t)|g_-; 0,1\rangle \right] \quad (2.45)$$

where $U(t)$ is the unitary time evolution operator of the coherent atom 2–field system. If the atom 2 velocity is so chosen that it receives a $\pi/2$ pulse in each cavity, $|\Psi_f\rangle$ becomes

$$|\Psi_f\rangle = \frac{1}{2}|e_+, g_-\,; 1, 0\rangle + \frac{1}{2\sqrt{2}}|e_+, e_+\,; 0, 0\rangle$$
$$- \frac{1}{2\sqrt{2}}|e_+, g_-\,; 0, 1\rangle + \frac{1}{2}|e_-, g_-\,; 0, 1\rangle$$
$$+ \frac{1}{2}|e_-, e_+\,; 0, 0\rangle. \tag{2.46}$$

One can check that in this state $\langle \sigma_{1x} \rangle = 0$. This implies that coherence is not restored by merely sending a second beam as described above through the cavities after the first beam. Had it been so, it would have been possible to send a superluminal signal by placing the Stern–Gerlach system far enough away downstream. However, suppose one detects atom 2 after it emerges from the cavities in state $|e_+\rangle$. Then there is a collapse of the state $|\Psi_f\rangle$ into the new state

$$|\Psi_f\rangle^{2 \text{ in } e_+} = \sqrt{\frac{2}{3}} \left[\frac{1}{\sqrt{2}}|e_+\rangle + |e_-\rangle \right]_1 \otimes |(e_+)_2\,; 0, 0\rangle, \tag{2.47}$$

which clearly shows that the states $|e_+\rangle$ and $|e_-\rangle$ of atom 1 are coherent. One now has

$$^{2 \text{ in } e_+}\langle \Psi_f | \sigma_{1x} | \Psi_f \rangle^{2 \text{ in } e_+} = \frac{\sqrt{2}}{3} \neq 0, \tag{2.48}$$

which shows that the interference is partially restored. Since $1/2 - \langle \sigma_{1x} \rangle = 0.0285$, the contrast in the pattern is quite high. The crucial point is the exploitation of *coincidence* counts in the Stern–Gerlach system for atom 1 and the *correlated* counts in the detector for atom 2 in a particular state. As this involves gathering information on both atoms 1 and 2 at the same time, it prevents superluminal signalling between them. In case atom 2 is detected in the other state $|g_-\rangle$, the imbalance in the detectors D_1 and D_2 is reversed. If these two alternatives are added up, the imbalance disappears and the interference is lost.

Quantum erasers have actually been demonstrated in quantum optics by using photon pairs generated by parametric down-conversion and polarization rotators as path detectors and erasers. Such erasers also display nonlocality and 'delayed choice' very dramatically. For further details and references to earlier work, see Herzog *et al.* [73].

3

Quantum nondemolition measurements

3.1   Introduction

Future high-precision measurements such as those being designed to
detect gravitational waves will involve the monitoring of a quantum
mechanical harmonic oscillator driven by a classical force. Typically,
a gravity wave antenna consists of an aluminium (or sapphire or silicon or niobium) bar weighing between 10 kilogrammes and 10 tonnes,
driven into oscillations of frequency $\omega/2\pi = 500$ to $10\,000$ hertz by
passing gravity waves. The type of gravity waves that are expected to
pass by the earth can cause only minute changes of oscillation amplitude $\delta x \simeq 10^{-19}$ cm, that are smaller than the quantum mechanical
wave packet $\delta x_{QM} = (\hbar/2m\omega)^{1/2}$ of the bar if the bar is in its ground
state or in a coherent state. *The macroscopic bar therefore behaves quantum
mechanically* as far as its degree of freedom corresponding to the amplitude
of oscillation is concerned. If one wants to see the details of the gravity
wave, one must be able to make repeated measurements of the bar's
amplitude with precision $\Delta x \leq 10^{-19}$ cm at time intervals of $\tau \leq 10^{-3}$ s
between successive measurements. In contrast to nuclear, atomic and particle physics, *there is only one quantum mechanical system rather than an
ensemble of systems on which a continual sequence of measurements must be
made.*

Such a bar will behave quantum mechanically even in the presence of
thermal Brownian motion and at temperatures $k_B T \gg \hbar\omega$ provided its
quality factor Q is sufficiently high, i.e. provided its fundamental vibra-
tional mode is weakly coupled to the thermalized modes.

If the bar is freely suspended like a pendulum, as it is in some detectors,
then over time intervals $\tau \sim 10^{-3}$ s it will behave as if it were not suspended
at all, and will move freely. Then, according to the Heisenberg uncertainty
relations, an 'initial' uncertainty $\Delta x_i \simeq 10^{-19}$ cm. will inevitably imply an

uncertainty of $\Delta p \geq (\hbar/2\Delta x_i)$ in its momentum, and hence an uncertainty $\Delta v = \Delta p/m \geq (\hbar/2m\Delta x_i)$ in its velocity, m being the mass of the bar. Between successive measurements the bar will then move away from its initial position by an amount $\Delta x_m = \Delta v\tau \geq (\hbar\tau/2m\Delta x_i) \geq 5 \times 10^{-19}$ cm, which is larger than the desired precision of the measurements. If the next measurement indicates a change in the position by as much as 5×10^{-19} cm, one cannot be certain whether the change was due to a passing gravity wave or to the unpredictable quantum mechanical uncertainty due to the previous measurement. The first measurement plus the subsequent free evolution of the bar will have 'demolished' all possibility of making a second measurement of the same precision as the first, and therefore of monitoring the bar and detecting the expected gravity wave.

One way of circumventing this problem would be to increase the mass of the antenna, but this is inconvenient. Another way would be to shorten the time between successive measurements, but this would weaken the gravitational wave more than it would reduce the quantum mechanical uncertainty in $\Delta x_m \propto \tau$.

A totally different approach being used is to strengthen the gravity wave effect by using laser interferometry. But this involves paying the price of measuring relative displacements of the order of $\sim 10^{-16}$ cm of two bars at a distance of several kilometres!

A completely novel approach is to circumvent the Heisenberg uncertainty relations by preventing the inevitable uncertainty arising from the first measurement from demolishing the possibility of a second accurate measurement – a quantum nondemolition (QND) measurement [74], [75], [76] For example, one could in principle measure the momentum of the bar with a small enough uncertainty, allow it to move freely for a short while and then measure the momentum again. Since only the position of the bar changes but not its momentum during free evolution, the uncertainty in the bar's position does not introduce any new uncertainty in its momentum which can therefore be repeatedly measured with the desired accuracy. This makes it possible to detect a momentum change larger than the initial uncertainty induced by a passing gravity wave. Thus, although position is not a QND variable, momentum is. Unfortunately, it is much more difficult to measure momentum than to measure position, and nobody has yet succeeded in inventing a technically feasible way of measuring momentum with the required accuracy.

The essential idea behind QND measurements is to use a measurement technique whose 'back-action' on the oscillator together with subsequent free evolution does not substantially disturb the probability distributions of the observables being measured. We will now describe the general theory of such QND measurements.

3.2   General theory of QND measurements

Consider some observable $\hat{A}$ of a quantum mechanical system (such as a harmonic oscillator) that an experimenter wishes to monitor. To start with let us assume that the only coupling of the system to the external world is through the measuring apparatus. Then a QND measurement of $\hat{A}$ is defined as a sequence of accurate measurements of $\hat{A}$ such that the result of every measurement is predictable from the result of the first measurement together with some information about the initial state of the system. The idea is to monitor the evolution of such an observable and watch for deviations from the predicted behaviour caused by a weak external force such as a gravity wave.

Not all observables can be monitored in a QND way. Suppose one is trying to monitor $\hat{A}$. Then the back-action of the measuring apparatus inevitably and uncontrollably affects all observables $\hat{C}$ that do not commute with $\hat{A}$. During the periods of free evolution between successive measurements these contaminations of $\hat{C}$ will feed back into $\hat{A}$, making the results of future measurements of $\hat{A}$ unpredictable. The surprising thing is that there *are* certain observables that are immune to such feedback contamination. They are called QND observables. The following is a mathematical criterion for a QND observable: an observable $\hat{A}$ is QND if and only if it commutes with itself at all instants of measurement t_i, t_j:

$$\left[\hat{A}(t_i), \hat{A}(t_j) \right] = 0. \tag{3.1}$$

If this condition is satisfied at all times t_i, t_j, $\hat{A}$ is called a *continuous* QND observable. This will obviously be the case if $\hat{A}$ is conserved during free evolution. If it holds at special times, it is called a *stroboscopic* QND observable.

As we have seen, for a free particle the momentum is conserved and is a continuous QND observable, as is also the energy. But the position is not, because $\hat{x}(t + \tau) = \hat{x}(t) + \hat{p}\tau/m$, so that

$$[\hat{x}(t), \hat{x}(t + \tau)] = i\hbar\tau/m. \tag{3.2}$$

This implies that precise measurements of $\hat{x}$ cause uncontrollable contaminations in $\hat{p}$, and these subsequently feed back into $\hat{x}$ as the particle moves freely.

The position and momentum operators of a harmonic oscillator satisfy the commutation relations

$$[\hat{x}(t), \hat{x}(t + \tau)] = (i\hbar/m\omega) \sin\omega\tau, \tag{3.3}$$

$$[\hat{p}(t), \hat{p}(t + \tau)] = (i\hbar m\omega) \sin\omega\tau. \tag{3.4}$$

Consequently, the position and momentum are not continuous QND variables. However, these commutation relations vanish for $\tau = k\pi/\omega$ (k = an integer), and so if measurements are made at time intervals of half periods or integral multiples thereof, position and momentum become stroboscopic QND variables.

The conserved quantities for a harmonic oscillator are its energy and the real and imaginary parts of its complex amplitude,

$$\hat{X}_1 = \hat{x}(t)\cos\omega t - [\hat{p}(t)/m\omega]\sin\omega t, \tag{3.5}$$

$$\hat{X}_2 = \hat{x}(t)\sin\omega t + [\hat{p}(t)/m\omega]\cos\omega t. \tag{3.6}$$

These are therefore continuous QND observables. Since

$$[\hat{X}_1, \hat{X}_2] = i\hbar/2m\omega, \tag{3.7}$$

one can make high-precision measurements of either $\hat{X}_1$ or $\hat{X}_2$ but not both. Such measurements are called 'back-action-evading' measurements, because they enable the measured component (say $\hat{X}_1$) to escape the uncontrollable back-action due to the measuring apparatus at the expense of contaminating the other component ($\hat{X}_2$).

In quantum mechanics the device used in any measurement consists of many stages. According to standard measurement theory the early stages of a measurement may be regarded as quantum mechanical, but the final stages must be classical. However, there is no universally accepted definition of what is classical. In other words, there is an inherent ambiguity in standard quantum mechanics regarding precisely where this 'cut' between the quantum mechanical and classical stages is to be placed. For pragmatic reasons it is usually assumed that a stage may be regarded as classical if the quantum mechanical uncertainties associated with it and of subsequent stages do not significantly affect the overall accuracy of the measurement. A measurement is called 'direct' if the system under study interacts directly with a classical apparatus as, for example, when the position of a particle is recorded by the blackening of a photographic plate. A measurement is called 'indirect' when between the system and the first classical stage there is a quantum mechanical stage (quantum mechanical readout system, QRS). An example is the measurement of the position of a particle by the scattering of light or electrons from it. The majority of measurements are indirect. In electronic measurements the first classical stage is often the first amplifier.

The fluctuations arising from the QRS and the measured system which are coupled must be very carefully estimated because they all influence the signal that enters the first classical stage, and it is this signal that determines the quantum limitations on the overall sensitivity of the measuring scheme under consideration.

The linearity of Schrödinger evolution ensures that when a measurement is made on a system S, it gets entangled with the measuring device M so that the state of the combined system $S + M$ is given by

$$|\psi\rangle = \sum_i c_i |S_i\rangle |M_i\rangle \tag{3.8}$$

where the index i denotes the complete set of orthogonal states of the system. The outcome of a particular result i indicated by by the corresponding state of the measuring device with the probability $|c_i|^2$ consists of a measurement. But such an outcome does not follow from Schrödinger evolution; it must be introduced as an additional assumption called the projection postulate. Such projections are nonlinear and non-unitary, and their effect on the state is to *reduce* it from a *pure* to a *mixed* state represented by the *reduced* density operator

$$\hat{\rho} = \sum_i \Pi_i \, \rho \, \Pi_i \tag{3.9}$$

where $\rho = |\psi\rangle\langle\psi|$ is the density operator corresponding to the pure state $|\psi\rangle$ and Π_i are projection operators ($\Pi_i^2 = \Pi_i$, $\sum_i \Pi_i = 1$). The question is: At which stage of the measurement does this reduction of the state occur? There is, as we have seen, no clear-cut answer to this in standard quantum mechanics. It is usually assumed that this reduction takes place when the signal from the QRS enters the so-called first classical stage defined above. If this signal contains information about both the QND observable $\hat{A}$ as well as other observables $\hat{C}$ which fail to commute with $\hat{A}$, then the back-action of the first classical stage on the earlier quantum stages will not only contaminate $\hat{C}$ but eventually also all observables that fail to commute with them, including $\hat{A}$, and no precise measurement of $\hat{A}$ will be possible. This shows that the final measurement error will always exceed the standard quantum limit (SQL). What is the standard quantum limit?

We will first deduce this SQL under the assumption that in the Heisenberg representation the observables $\hat{A}$ and $\hat{C}$ are time-independent, i.e. they are either constants of motion like $\hat{X}_1$ or $\hat{X}_2$ or are evaluated at a given moment in time (such as $\hat{A} = \hat{x}(0)$ and $\hat{C} = \hat{p}(0)$). Let $\hat{Q}_R$, the readout observable from the QRS that enters the first classical stage, be expressed by

$$\hat{Q}_R = f(\alpha\hat{A} + \beta\hat{C}) \tag{3.10}$$

where

$$[\hat{A}, \hat{C}] = 2i\gamma\hbar \tag{3.11}$$

so that we have the uncertainty relation

$$\Delta\hat{A} \, \Delta\hat{C} \geq \gamma\hbar. \tag{3.12}$$

The function f as well as the real parameters α and β contain information about the time evolution of $\hat{Q}_R$. Typically, α and β vary sinusoidally with time, and can be used to code and separate the $\hat{A}$ and $\hat{C}$ signals. If we assume that the first classical stage is equally sensitive to the $\hat{A}$ and $\hat{C}$ frequencies, then no matter how precisely it monitors $\hat{Q}_R$, it is bound to introduce errors in $\hat{A}$ and $\hat{C}$ related by $\Delta\hat{A} = (\tilde{\beta}/\tilde{\alpha})\Delta\hat{C}$ where $\tilde{\alpha}$ and $\tilde{\beta}$ are the rms values of α and β. Combined with the uncertainty relation (3.12) this gives the SQL

$$\Delta\hat{A} \geq \left[(\tilde{\beta}/\tilde{\alpha})\gamma\,\hbar\right]^{1/2}. \tag{3.13}$$

This shows that $\hat{A}$ can be measured precisely only if β is zero.

Let us now see what happens if $\hat{A}$ and $\hat{C}$ are time-dependent. In order to ensure that the instantaneous QRS that enters the first classical stage does not contain any contamination information about the observables $\hat{C}(t)$, it is necessary and sufficient that $\hat{A}(t)$ commute with the piece of the Hamiltonian $\hat{H}_I(t)$ that describes the interaction of the system with the measuring device:

$$[\hat{A}(t), \hat{H}_I(t)] = 0. \tag{3.14}$$

This guarantees that there is no instantaneous back-action of the measuring device on $\hat{A}(t)$. One has also to make sure that the observables that have been contaminated by back-action of the measuring device do not subsequently, during the course of free evolution when $\hat{H}_I(t)$ is turned off, contaminate $\hat{A}(t)$. This can be arranged by direct coupling of $\hat{A}(t)$ to some observable $\hat{M}$ of the measuring device:

$$\hat{H}_I(t) = K\hat{A}(t)\hat{M}. \tag{3.15}$$

However, sometimes H_I has to be kept turned on for a long time, and there is then the possibility that H_I may cause indirect evolutionary contamination of $\hat{A}(t)$ through $\hat{C}(t)$. Fortunately it can be shown [76] that $\hat{A}(t)$ is fully immune from all kinds of back-action contamination provided it is a continuous QND observable and H_I does not contain any system observable other than $\hat{A}(t)$.

Summarizing the above discussion, we can say that a measurement is of the QND type if the observable $\hat{A}$ to be monitored, the interaction Hamiltonian $\hat{H}_I$ between the system and the measuring apparatus and the readout observable $\hat{Q}_R$ satisfy the following conditions:

(1) $\hat{H}_I$ is a function of $\hat{A}$,

(2) $\left[\hat{H}_I, \hat{A}\right] = 0$,

(3) $\left[\hat{H}_I, \hat{Q}_R\right] \neq 0$, and

(4) $\left[\hat{A}(t_i), \hat{A}(t_j) \right] = 0.$

Condition (4) is satisfied if the free Hamiltonian $\hat{H}_S$ of the system does not contain the conjugate observables $\hat{C}$ of $\hat{A}$ with which it does not commute.

We are now in a position to apply the theory of QND measurements to monitor an external classical force $F(t)$. Let the system be coupled to $F(t)$ through the Hamiltonian

$$\hat{H}_F = \mu F(t) \hat{O}(t) \qquad (3.16)$$

where μ is a coupling constant and $\hat{O}$ is a suitable dynamical variable of the system. (In the case of a gravitational antenna $\hat{O}$ is the position $\hat{x}$.) The procedure is to make a sequence of measurements of a QND observable $\hat{A}(t)$ of a system and to detect the force by the changes it produces in the precisely predictable eigenvalues $A(t)$ of $\hat{A}(t)$ which would be measured in the absence of the force. It follows from the above discussion that in order to produce instantaneous and arbitrarily precise measurements of $\hat{A}(t)$ in the presence of the external force, $\hat{A}(t)$ must be a continuous QND observable in the presence of $F(t)$. Further, the eigenvalues $A(t)$ of $\hat{A}$ must evolve in such a way in the presence of $\hat{H}_F$ that

$$A(t) = f \left[A_0; F(t'); t, t_0 \right] \quad \text{for} \quad t_0 < t' < t \qquad (3.17)$$

is a uniquely invertible functional of $F(t')$ for every eigenvalue A_0 that is a possible result of the first measurement of $\hat{A}(t_0)$. It is then possible to deduce $F(t)$ from a precise knowledge of $A(t)$, and $\hat{A}$ is called a QNDF variable. (Unruh [75] refers to them as QNDD variables.)

In the case of a gravitational antenna with position coupled to the force, the real and imaginary parts $\hat{X}_1$ and $\hat{X}_2$ of the complex amplitude (3.5) and (3.6) are QNDF observables and can be used to monitor the force. However, although the energy is a QND observable, it is not QNDF. Consequently, although it can be used to detect the presence of an arbitrarily weak external force, it cannot be used to measure its strength with a precision better than a factor of 3 [76] unless the force is so strong that it increases the energy by an amount large compared to the initial energy.

Since the amplitude and phase of an oscillator are non-commuting observables, they cannot be precisely measured simultaneously. Suppose an oscillator is being monitored in the standard way by a sequence of 'amplitude-and-phase' measurements, each of duration $\bar{\tau} \gg 2\pi/\omega$. Then one can show [77] that an external driving force $F = F_0 \cos(\omega t + \phi)$ can be detected provided F_0 exceeds a certain quantum limit (see (3.32)). No standard 'amplitude-and-phase' measurement can do better than that.

There are two ways of beating this limit: (a) the 'back-action-evading' measurements of the real part $\hat{X}_1$ (or the imaginary part $\hat{X}_2$) of the complex amplitude which can lead to an arbitrarily accurate monitoring of the classical force; and (b) the QND technique of 'quantum counting' which can detect an arbitrarily weak classical force but cannot provide good accuracy in determining its precise time dependence. We will now discuss in greater detail how these schemes can be carried out in principle.

3.2.1 The back-action-evading method

Caves *et al.* [76] have shown that non-relativistic quantum mechanics permits 'arbitrarily quick and accurate' measurements of $\hat{X}_1$ provided (i) the measuring apparatus is coupled to $\hat{X}_1$ and no other oscillator observable and (ii) the coupling between the measuring apparatus and the oscillator is arbitrarily strong. We have already seen that condition (i) ensures complete shielding of $\hat{X}_1$ from the back-action of the measuring apparatus, while condition (ii) leads to arbitrarily good accuracy for any measurement time, however short. (In practice, however, the strengths of real materials, voltage breakdown in capacitors, etc. will limit the quickness and precision with which $\hat{X}_1$ can be measured.)

For the time being let us assume that a suitable arrangement for such quick and accurate measurements has been set up, and that an initial precise measurement of $\hat{X}_1$ at time t_0 puts the oscillator in an eigenstate $|\xi_0\rangle$ with eigenvalue ξ_0 which can be any real number because $\hat{X}_1$ has a continuous spectrum of eigenvalues. The state of the system remains fixed in time in the Heisenberg picture at $|\xi_0\rangle$, but if the total Hamiltonian is

$$\hat{H} = \hat{H}_0 - \hat{x}\,F(t), \tag{3.18}$$

$$\hat{H}_0 = \frac{\hat{p}^2}{2m} + \frac{1}{2}m\omega^2\hat{x}^2, \tag{3.19}$$

$\hat{X}_1$ evolves:

$$\frac{\mathrm{d}\hat{X}_1}{\mathrm{d}t} = -\frac{i}{\hbar}[\hat{X}_1, \hat{H}] + \frac{\partial \hat{X}_1}{\partial t} = -\frac{F(t)}{m\omega}\sin\omega t. \tag{3.20}$$

Integrating this equation, one obtains

$$\hat{X}_1 = \hat{X}_1(t_0) - \int_{t_0}^{t} \frac{F(t')}{m\omega}\sin(\omega t')\,\mathrm{d}t'. \tag{3.21}$$

Since $|\xi_0\rangle$ is an eigenstate of $\hat{X}_1(t_0)$ and the integral in the above equation is a real number, $|\xi_0\rangle$ is also an eigenstate of $\hat{X}_1(t)$ with eigenvalue

$$\xi(t, t_0) = \xi_0 - \int_{t_0}^{t} \frac{F(t')}{m\omega}\sin(\omega t')\,\mathrm{d}t'. \tag{3.22}$$

A precise measurement of $\hat{X}_1(t)$ at any time t must give this eigenvalue and leave the oscillator state unchanged (except for overall phase). This elucidates why this is called a back-action-evading *nondemolition* measurement. One can therefore monitor, in principle, the precise time evolution of the oscillator's eigenvalue $\xi(t, t_0)$ by a sequence of arbitrarily quick and precise measurements of $\hat{X}_1$, and from it compute the precise time evolution of the driving force (signal):

$$F(t) = -(m\omega \mathrm{d}\xi/\mathrm{d}t)/\sin\omega t. \tag{3.23}$$

In the realistic case, of course, the measurements will not be absolutely precise, and the measured values of $F(t)$ will be specially inaccurate at times $t \simeq n\pi/\omega$ when $\sin\omega t \simeq 0$. However, if the driving force $F(t)$ is *classical* and has a long enough wavelength that is larger than the size of the measuring apparatus, one can use a second oscillator coupled to $F(t)$. One has to measure $\hat{X}_1$ on the first oscillator to get $\xi(t, t_0)$ and $\hat{X}_2$ (the imaginary part) on the second to get the corresponding eigenvalues

$$\zeta(t, t_0) = \zeta_0 + \int_{t_0}^{t} \frac{F(t')}{m\omega} \cos(\omega t')\,\mathrm{d}t'. \tag{3.24}$$

From these one can independently infer $F(t)$, and the accuracies of the two sets of measurements will be complementary: the first will be accurate at times $t = (n + \frac{1}{2})\pi/\omega$ when the second is poor, and the second will be accurate at times $t = n\pi/\omega$ when the first is poor. In this way the uncertainty relations can be completely circumvented, and the system point in the complex amplitude plane $\hat{X}_1$–$\hat{X}_2$ will evolve in exactly the same manner as X_1 would evolve for a single *classical* oscillator driven by $F(t)$.

We will now briefly discuss how a back-action-evading precise measurement of $\hat{X}_1$ can be arranged. The constraints (i) and (ii) mentioned at the beginning of this section force the interaction Hamiltonian H_I between the oscillator and the measuring apparatus to be of the form

$$\hat{H}_I = K\hat{X}_1\hat{Q} = K\left[\hat{x}\cos\omega t - (\hat{p}/m\omega)\sin\omega t\right]\hat{Q} \tag{3.25}$$

where K is a coupling constant that may be time-dependent, and $\hat{Q}$ is an observable of the measuring apparatus which commutes with all the oscillator observables. Specific designs of this type of measuring apparatus have been described by Caves *et al.* [76] and will not be discussed here. Suffice it to say that the most promising method for gravitational wave detection appears to be *continuous single-transducer measurements* (using a single transducer with sinusoidally modulated coupling to the oscillator, followed by a filter which averages the transducer output over a time $\bar{\tau} \gg 2\pi/\omega$). Such measurements would require modest modifications of the amplitude-and-phase electronic techniques already in use.

3.2.2 The quantum counting method

The free Hamiltonian of an oscillator is generally of the form

$$H_S \equiv \hat{p}^2/2m + \frac{1}{2}m\omega^2\hat{x}^2, \tag{3.26}$$

and the number operator is given by

$$\hat{N} \equiv \hat{A}^\dagger\hat{A} = \hat{H}_S/\hbar\omega - \frac{1}{2} \tag{3.27}$$

where $\hat{A}$ and $\hat{A}^\dagger$ are the creation and annihilation operators. Since the real and imaginary parts of the complex amplitude are given by (3.5) and (3.6), one can also express $\hat{N}$ as

$$\hat{N} = \frac{1}{2}(m\omega/\hbar)\left(\hat{X}_1^2 + \hat{X}_2^2\right) - \frac{1}{2}. \tag{3.28}$$

Hence a precise measurement of the absolute amplitude $|X| = (X_1^2 + X_2^2)^{1/2}$ of the oscillator is equivalent to a measurement of $\hat{N}$. However, such a measurement leaves the phase of the oscillator $\psi = \tan^{-1}(X_2/X_1)$ completely undetermined.

Quantum counters with high efficiencies are available for photons of optical frequencies and higher, but they destroy the photons by absorbing them; they are demolition devices. QND methods of counting photons in a microwave cavity were first proposed by Braginsky, Vorontsov and Khalili [74] and Unruh [75]. Braginsky and Vorontsov [78] also suggested that such a cavity could be coupled to a resonant bar in such a fashion that the bar phonons are converted into cavity photons which can then be counted by a QND method, providing a technique to monitor the changes in the number of bar phonons.

Let us now see how to monitor a classical force acting on the oscillator. Consider a precise measurement of the number of quanta at time $t = 0$ that puts the oscillator into an energy and number eigenstate $|N\rangle$ with energy E_0 and N_0 quanta. This state will change if a force $F_0\cos(\omega t + \phi)$ acts on the oscillator for a time τ. Since the initial phase of the oscillator is completely indeterminate, the interference terms in the expectation value of the energy of the new state drop out, and one has [79]

$$\langle E \rangle = E_0 + W, \tag{3.29}$$
$$W = F_0^2\tau^2/8m. \tag{3.30}$$

However, interference plays a dominant role in the variance of the new state's energy:

$$\sigma(E) = (2 E_0 W)^{1/2}. \tag{3.31}$$

If the force is weak ($W < E_0$), the next measurement can reveal its presence provided $\sigma(E) \geq \hbar\omega$. Together with (3.30) and (3.27) this implies

$$F_0 \geq \frac{2}{\tau} \left(\frac{m\omega\hbar}{N_0 + \frac{1}{2}} \right)^{1/2}. \tag{3.32}$$

If N_0 is made arbitrarily large, this detection method can be arbitrarily sensitive. However, since $\sigma(E) \gg \langle E \rangle - E_0$, there is no unique relationship between the measured energy and F_0, and this method fails to give us the precise magnitude of F_0 – the energy is not a QNDF observable. (For different derivations of the detection criterion (3.32), see Caves *et al.* [76] and Hollenhorst [80].)

Discussions on coupling the oscillator's energy to a QRS will be found in Braginsky, Vorontsov and Khalili [74], Unruh [75] and Braginsky, Vorontsov and Thorne [77].

More recent QND methods have been proposed by Milburn and Walls [81], Imoto, Haus and Yamamoto [82], Levenson, Shelby, Reid and Walls [83], La Porta, Slusher and Yurke [84], Brune, Haroche, Lefevre, Raimond and Zagury [60] and Grangier, Roch and Roger [85].

4

Topological phases

4.1 The Aharonov–Bohm (AB) effect

In classical electrodynamics the scalar and vector potentials $(\Phi, \vec{A})$ have no direct physical significance. They are merely used as auxiliary variables that simplify computations of the physical fields $F_{\mu\nu}$. This is because the fields, defined in terms of the potentials by $F_{\mu\nu} = \partial_\mu A_\nu - \partial_\nu A_\mu$, are invariant under the gauge transformations $A_\mu \rightarrow A_\mu + \partial_\mu \chi$ on the potentials, where χ is an arbitrary function of space–time. The fields therefore do not uniquely fix the potentials. In quantum mechanics, however, the concept of force fields fades away and the potentials themselves acquire a direct physical significance, as was first clearly pointed out by Aharonov and Bohm [86] based on an earlier suggestion by Ehrenberg and Siday[87].

Consider a double-slit interference experiment with electrons in which an infinitely long cylindrical impenetrable solenoid of diameter small in comparison with the distance between the slits, is placed just behind the screen with the two slits (Fig. 4.1). In such a configuration the probability that the electrons diffracted at the slits will get near the solenoid is negligible. When a current passes through the solenoid, it generates a magnetic field $\vec{B}$ inside the solenoid but none outside, though circulating potentials $\vec{A}$ appear all around the solenoid in the region R accessible to the electrons. Thus, although there is no magnetic field in the region R, the potentials there will still influence the electrons according to quantum mechanics, resulting in an *additional* phase shift given by

$$\frac{q}{\hbar} \oint \vec{A} \cdot d\vec{l} = \frac{q}{\hbar} \int \text{curl} \vec{A} \cdot d\vec{s} = \frac{q}{\hbar} F, \qquad (4.1)$$

where the integral is around any closed path encircling the solenoid and F is the flux of $\vec{B}$ through it. This is the famous Aharonov–Bohm effect. Classically such an effect cannot occur because the classical Lorentz force on an electron depends on the local magnetic field $\vec{B}$ which is zero

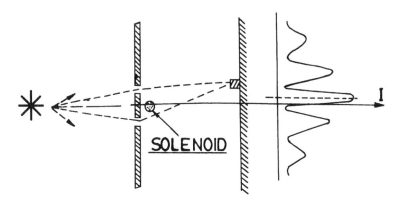

Fig. 4.1. A modified double-slit experiment illustrating the Aharonov–Bohm effect.

everywhere in R. Quantum mechanically, however, one can tell that there is a magnetic field in the *inaccessible* region inside the solenoid by just going around it without ever going close to it. This a kind of nonlocal effect that a local classical field cannot have. The potentials therefore acquire a new physical significance in quantum mechanics that is at first hard to accept.

Notice that the closed line integral is actually gauge invariant, because the additional term due to the gauge transformation is the gradient $\nabla \chi$, and the tangential component of a gradient on a closed path is always zero by Stokes' theorem. Thus, although the potential has a non-classical physical influence, the physical effect is still gauge invariant as in classical electrodynamics. Consider the form of the vector potential in the field-free region R outside the cylindrical solenoid expressed in cylindrical coordinates ρ, z, ϕ (with the z-axis along the axis of the solenoid),

$$A_\rho = A_z = 0, \quad A_\phi = D/2\pi\rho, \tag{4.2}$$

where D is any constant independent of the spatial coordinates. This choice satisfies the condition curl $\vec{A} = 0$. In order that Stokes' theorem holds, namely

$$\oint \vec{A} \cdot \mathrm{d}\vec{l} = \int \mathrm{curl}\, \vec{A} \cdot \mathrm{d}\vec{s} = \frac{\hbar}{q} F, \tag{4.3}$$

one must have

$$A_\rho = A_z = 0, \quad A_\phi = F/2\pi\rho. \tag{4.4}$$

Inside the solenoid the vector potential reproducing the uniform field **B** must be of the form

$$A_\rho = A_z = 0, \quad A_\phi = B\rho/2. \tag{4.5}$$

Outside the solenoid, it might appear that one can choose the potential to be

$$A_\rho = A_z = A_\phi = 0. \tag{4.6}$$

In such a non-Stokesian gauge the AB effect obviously does not exist. Since the magnetic field may not be continuous across the boundary of the solenoid, it is not clear that Stokes' theorem can be applied to this case. Thus, it appears that the two apparently equivalent gauges (4.3) and (4.6) give rise to different physical effects. This is puzzling, and was the source of some early confusion regarding the reality of the AB effect [88].

The confusion can be resolved by taking into account the dynamical nature of the process of buildup of the magnetic field inside the solenoid and of the vector potential in R as the current is switched on [89]. In such a transient situation the flux F varies with time. If the boundary of the solenoid (or finite magnetic whisker in real situations) is not impenetrable to the electric field $\vec{E}$, the electric field induced by the time variation of F will be continuous across the boundary of the solenoid. Hence Stokes' theorem *can* be applied to $\vec{E}$. It follows from the field equation $\mathrm{curl}\,\vec{E} = -(1/c)\,\partial\vec{B}/\partial t$ that for any arbitrary closed curve in the field-free region R,

$$\oint \vec{E} \cdot d\vec{l} = -\frac{1}{c}\frac{d\,F}{d\,t}. \tag{4.7}$$

Using $\vec{E} = -(1/c)(\partial\vec{A}/\partial t)$ and taking the scalar potential $V = 0$, one obtains

$$\oint \vec{E} \cdot d\vec{l} = -\frac{1}{c}\frac{\partial}{\partial t}\oint \vec{A} \cdot d\vec{l}. \tag{4.8}$$

If one uses the choice of gauge (4.2), one obtains

$$\oint \vec{E} \cdot d\vec{l} = -\frac{1}{c}\frac{dD}{dt}. \tag{4.9}$$

Comparing (4.7) and (4.9), one obtains

$$D(t) = F(t) + F_0 \tag{4.10}$$

where F_0 is an arbitrary constant independent of time. This is the most general restrictive condition on the gauge parameter D that can be derived from the basic classical field equations. Note that when $F = 0$, $\vec{B}$ must be zero both inside and outside the solenoid, and so $\vec{B}$ is continuous across the boundary of the solenoid. Stokes' theorem can therefore be applied, leading to

$$\oint \vec{A} \cdot d\vec{l} = 0, \tag{4.11}$$

which implies $D = 0$ when $F = 0$. It follows therefore from the relation
(4.10) that $F_0 = 0$. This uniquely fixes the vector potential at an arbitrary
instant in the field-free region R and the choice of gauge given by (4.4).
It turns out that if Stokes' theorem is valid at any instant and the electric
field is continuous, the field equations guarantee that the theorem is valid
at all instants. Thus, although it might appear in the static situation that
the choice of gauge (4.6) is possible, it turns out to be inconsistent with
the time-dependent buildup process of the field $\vec{B}$ inside the solenoid and
of the vector potential $\vec{A}$ in the region R. This resolves the confusion and
establishes the existence of the AB effect.

In fact, that the existence of the AB effect is actually necessary for the
complementarity of path distinguishability and interference of a charged
particle had already been clearly demonstrated by Furry and Ramsey [90].

Wu and Yang [91] have given an exposition of the AB effect that
underscores the fact that the field strengths alone underdescribe electromagnetism. They showed that in a multiply connected region such as R
where the fields vanish everywhere, the physical AB effect is determined
by the non-integrable *phase factor*

$$\exp\left[\frac{q}{\hbar} \oint A_\mu \, dx^\mu\right] \qquad (4.12)$$

around an unshrinkable loop and not the phase

$$\oint A_\mu dx^\mu. \qquad (4.13)$$

The phase (4.13) contains more information than the *phase factor* (4.12)
but this extra information has no physical significance. This is in contrast
to the claim that electromagnetism can be described in terms of field
strengths alone [92]. They also extended the AB effect to non-Abelian
gauge fields such as SU_2.

4.2   Experimental verification of the AB effect

A number of experiments were performed which claimed to have verified
the AB effect [93]. In all these experiments the solenoids or magnetic
whiskers used were of finite length. It was pointed out by Roy [94] that
none of these experiments could have established the AB effect (essentially
an effect of inaccessible fields) because, with impenetrable solenoids or
whiskers of *finite* length, the effects of a vector potential on a charged
particle in the accessible region R are completely determined by the field
strengths in R alone. The crux of the idea is that with a solenoid or
whisker of finite length there is inevitably some leakage of the magnetic
flux through the ends of the solenoid to the accessible region R, and the

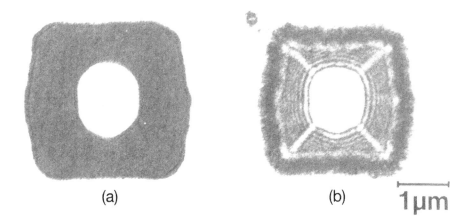

Fig. 4.2. Toroidal magnet: (a) electron-micrograph image (b) Lorentz micrograph (after Ref. [95]).

continuity of the magnetic field in such cases and the equation $\mathrm{div}\vec{B} = 0$ guarantees

$$\oint_C \vec{A} \cdot \mathrm{d}\vec{l} = \int_S \vec{B} \cdot \mathrm{d}\vec{s} = F, \qquad (4.14)$$

where the line integral is over any closed curve C in R and S is any arbitrary surface lying wholly in R with the closed curve as its boundary. This shows that with finite solenoids $\oint \vec{A} \cdot \mathrm{d}\vec{l}$ can be expressed wholly in terms of the local magnetic field in the region R alone.

The only way of avoiding flux leakage is to use a magnetic field completely confined within a toroidal magnet. Such an experiment was performed by Tonomura *et al.* [95] who fabricated tiny circular toroidal ferromagnets using electron-beam lithography. Electron microscopic images and underfocussed Lorentz micrographs were taken (Fig. 4.2). The latter showed that there was no appreciable flux leakage. Using electron holographic interferometry, they claimed to have established the AB effect.

The experimental arrangement was such that the electron beam partially touched and even penetrated the magnet. This point is open to criticism. They, however, argued that since the shape of the magnetic sample was reproduced as a clear image on the interferogram, the part of the beam that was transmitted through the magnetic flux in the sample could not have contributed to points outside the sample image. The beams reaching such points could have 'felt' only the magnetic vector potential, if any. Moreover, different electron energies (80, 100 and 125 kV) were used to achieve different penetrations, but no change in

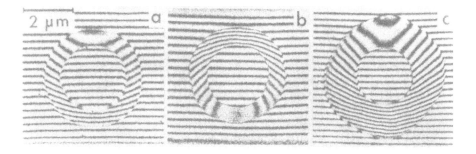

Fig. 4.3. Interferograms of toroidal magnets. (a) $\Phi = 1.2\,(h/e)$. (b) $\Phi = 2.0\,(h/e)$. (c) $\Phi = 2.8\,(h/e)$. Fringes are, in general, not on the same straight line in the two spaces inside and outside a toroid, except for accidental coincidence (b) (after Ref. [97]).

the phase difference was noticed. Regardless of the strength of penetrability, the experimentally obtained interference patterns could be fully explained in terms of the Stokes potential, (4.4). This could not be expected had the vector potential been everywhere zero outside the toroid, (4.6).

Subsequently, it was pointed out [96] that the possible quantization of the magnetic flux in $h/2e$ units in a rigorously impenetrable toroidal magnet (an 'autistic magnet') would result in a drastic change in the interpretation of the AB effect [96]. Quantization results if one assumes single-valuedness of the electron wave function, which Aharonov and Bohm did not. Also, there is some unavoidable electron tunneling across the magnet's boundary in such a situation, rendering their impenetrability assumption invalid. This means that the effect does not really exist if the flux is quantized. This could not be verified in the above experiment because the leakage flux produced had an error of $\pm h/2e$ in the flux measurement. The experiment was therefore repeated with phase amplification to obtain greater accuracy [97]. The results showed that the magnetic flux in a toroidal magnet is not quantized and has a continuous value within an experimental error of $\pm h/10e$, as is evident from the interferograms which show that the interference fringes in the spaces inside and outside the toroid are not on the same straight lines (Fig. 4.3) as would have been the case had the trapped flux been quantized. This is also evidence against the single-valuedness of the electron wave function. Indeed, there is no fundamental reason in quantum mechanics for it to be so. The principle of the hologram formation is shown in Fig. 4.4. Figure 4.5 shows the principle of the optical reconstruction for interference microscopy carried out with a He–Ne laser of wavelength 6328 Å.

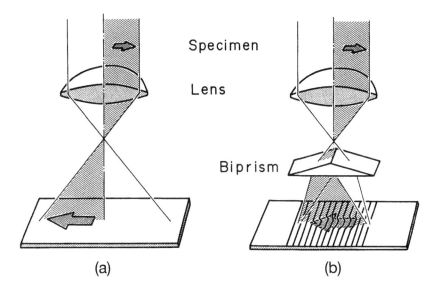

Fig. 4.4. Sketch of hologram formation with an electron biprism. (a) Electron microscopic image. (b) Off-axis electron hologram (after Ref. [97]).

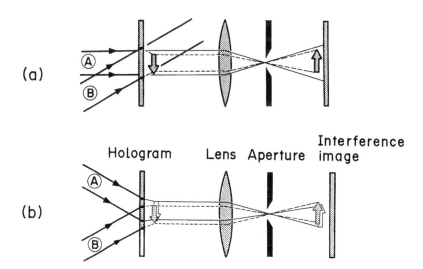

Fig. 4.5. Sketch of optical reconstruction for interference microscopy. (a) Normal interference micrograph. (b) Doubly amplified interference micrograph (after Ref. [97]).

4.3   The scalar AB effect

In their original paper Aharonov and Bohm had pointed out two types of effects, the usual magnetic (or vector) AB effect, and the less often cited electric (or scalar) AB effect caused by the scalar potential $V = -eU$ in the Schrödinger equation. Figure 4.6(a) shows a divided electron wave packet travelling down to two cylinders which act as Faraday cages at potentials U_1 and U_2 whose interiors are field-free. To observe the scalar AB effect, only one of the cylinders (say cylinder 2) is pulsed during the times when there is an electron wave packet inside it. In spite of there being no electric field inside at any time, a relative phase shift $\Delta\phi$ is predicted:

$$\Delta\phi = \frac{1}{\hbar} \int e\, U_2(t)\, \mathrm{d}t\,. \qquad (4.15)$$

An experiment with neutrons was actually carried out by Allman *et al.* [98]. The phase shift was produced by a scalar potential $V = -\vec{\mu}.\vec{B}$ which is the analogue of $V = -eU$. A current pulse $i_2(t)$ is applied to the solenoid 2 (Fig. 4.6(b)) when there is a neutron wave packet inside it. The magnetic field $\vec{B}_2(t)$ gives rise to the relative phase shift

$$\Delta\phi_{AB} = (\sigma/\hbar) \int \mu B_2(t)\, \mathrm{d}t \qquad (4.16)$$

where $\sigma = \pm 1$ depending on whether the neutron is spin up or spin down relative to the magnetic field, the direction of quantization. In the actual experiment short current pulses of duration $8\,\mu s$ were applied to a suitably designed solenoid placed inside a skew symmetric single-crystal neutron interferometer (Fig. 4.7). The pulses were applied cyclically and each detected neutron was gated into a separate scalar according to its arrival time within the cycle. In this way the phase shift of neutrons that traversed the solenoids when the current was zero could be compared with the phase shift of the neutrons that traversed the solenoids when the current, and hence the magnetic field, was non-zero. By adjusting both the orientation of the Al plate and the d.c. current in the bias coil, it was possible separately to control the phase shifts for the spin up and spin down states. With these appropriate phases, it was possible to determine the AB phase shift $\Delta\phi_{AB}$ with maximum sensitivity, even using unpolarized beams. The difference in counts between the positive and negative polarity pulses recorded by the detector C_3 is predicted to be

$$N_3(-) - N_3(+) = 2\,N_1\, b_3 \sin(\Delta\,\phi_{AB}), \qquad (4.17)$$

which is in clear agreement with the data (Fig. 4.8).

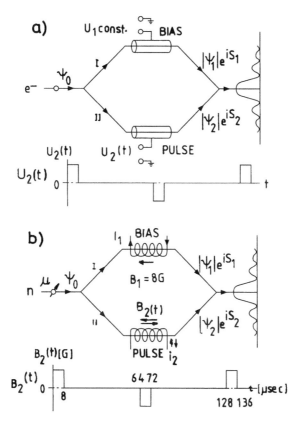

Fig. 4.6. Schematic diagram of the scalar AB experiment for (a) electrons and (b) neutrons. The wave forms of the applied pulses are also shown (after Ref. [98]).

4.4 The Anandan–Aharonov–Casher (AAC) effect

That neutral particles with an anomalous magnetic moment should exhibit the 'dual' of the magnetic or vector AB effect in an electric field was first pointed out by Anandan [99] and later independently by Aharonov and Casher [100]. The Lagrangian density of an electrically neutral Dirac particle with a magnetic moment of mass M (such as a neutron) is

$$\mathscr{L} = \bar{\psi} \left[i\gamma^\mu \, \partial_\mu - M - \frac{1}{2} \mu F^{\mu\nu} \sigma_{\mu\nu} \right] \psi. \tag{4.18}$$

Using the representation

$$\gamma^0 = \begin{pmatrix} 1 & 0 \\ 0 & 1 \end{pmatrix}, \quad \sigma^k = \begin{pmatrix} 0 & \sigma^k \\ \sigma^k & 0 \end{pmatrix},$$

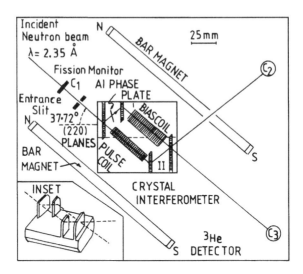

Fig. 4.7.   Layout of the scalar AB experiment using a skew-symmetric Si single-crystal neutron interferometer. Inset: an isometric view of the interferometer crystal (after Ref. [98]).

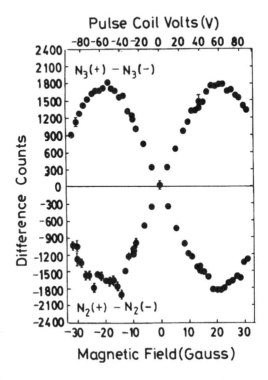

Fig. 4.8.   Interferometer output signals as a function of pulse-coil field strengths (after Ref. [98]).

one can easily find in the non-relativistic limit the effective Hamiltonian for a magnetic moment in an external electric field $\vec{E}$:

$$
\begin{aligned}
H_{\text{NR}} &= \frac{1}{2M} \vec{\sigma} \cdot (\vec{p} - i\mu\vec{E}) \, \vec{\sigma} \cdot (\vec{p} + i\mu\vec{E}) \\
&= \frac{1}{2M} (\vec{p} - \vec{E} \times \vec{\mu})^2 - \frac{1}{M} \mu^2 E^2
\end{aligned}
\tag{4.19}
$$

where $\vec{\mu} = \mu\vec{\sigma}$. If $\vec{E}$ is the Coulomb field of a charged particle, then

$$
\vec{E} \times \vec{\mu} = \frac{e}{4\pi} \frac{(\vec{R} - \vec{r}) \times \vec{\mu}}{|\vec{R} - \vec{r}|^3} = e\vec{A}(\vec{r} - \vec{R}),
\tag{4.20}
$$

where $\vec{A}$ is the vector potential at the point $\vec{r}$ with the magnetic moment $\vec{\mu}$ at the point $\vec{R}$ as its source! Thus, the non-relativistic Lagrangian corresponding to the Hamiltonian (4.19) is

$$
L = \frac{1}{2} MV^2 - e\vec{A}(\vec{r} - \vec{R}) \cdot \vec{V},
\tag{4.21}
$$

neglecting terms of order $\mathcal{O}(\mu^2 E^2)$. Adding to this the Lagrangian of an electrically charged particle of mass m at $\vec{r}$ interacting with a solenoid located at the point $\vec{R}$ (in a plane), we obtain the Lagrangian for a system of a charged particle interacting with a magnetic moment:

$$
L = \frac{1}{2} mv^2 + \frac{1}{2} MV^2 + e\vec{A}(\vec{r} - \vec{R}) \cdot (\vec{v} - \vec{V}).
\tag{4.22}
$$

This implies a gauge-invariant force that vanishes with the magnetic field:

$$
\begin{aligned}
m\dot{\vec{v}} &= \frac{1}{2} e(\vec{v} - \vec{V}) \times \vec{B} = -M\dot{\vec{V}} ; \\
\vec{B} &= \vec{\nabla} \times \vec{A}(\vec{r} - \vec{R}).
\end{aligned}
\tag{4.23}
\tag{4.24}
$$

Because of the dependence of $\vec{A}$ on $(\vec{r} - \vec{R})$, the system possesses a kind of duality when the roles of $\vec{r}$ and $\vec{R}$ are reversed. Although a magnetic moment moving with its moment parallel to a straight homogeneously charged line feels no classical force, it undergoes a phase shift given by

$$
\begin{aligned}
S_{\text{AAC}} &= -\oint e\vec{A}(\vec{r} - \vec{R}) \cdot d\vec{R} = \mu\lambda \\
&= 2\pi\alpha g/M\xi
\end{aligned}
\tag{4.25}
$$

where λ is the charge per unit length on the line, μ the projection of the magnetic moment along the line, $\xi = e/\lambda$, $\mu = ge/2M$ with g as the g-factor and α is the fine-structure constant. For a neutron, $M^{-1} = 2 \times 10^{-14}$ cm, while for an atom $M^{-1} \sim 4 \times 10^{-11}$ cm because the mass scale is controlled by the electron mass. Therefore, in order to observe the AAC effect ($S_{AAC} \sim \pi/2$) one requires linear charge densities

$$\text{neutron}: \quad \lambda \sim e/10^{-15}\,\text{cm},$$

$$\text{atom}: \quad \lambda \sim e/2 \times 10^{-12}\,\text{cm} \qquad (4.26)$$

Although these densities are huge, the effect should be observable with neutrons and/or atomic beams since there is no limitation in principle on the thickness of the charged line.

If the solenoid moves arbitrarily, it produces an electric as well as a magnetic field. The corresponding AB phase shift for a charged particle can be expressed covariantly in terms of the electric and magnetic fields. The dual of this is the AAC phase shift produced by the magnetic and electric fields of a line charge in arbitrary motion acting on a neutral particle with an anomalous magnetic moment. To lowest order in the WKB approximation, the normalized wave function corresponding to (4.18) is [101]

$$\psi(x) = \exp\left(-\mathrm{i}\int_\gamma p_\mu \mathrm{d}x^\mu\right) \times P \exp\left(-\mathrm{i}\mu \int_\gamma a_\mu \mathrm{d}x^\mu\right)\psi(x_0) \qquad (4.27)$$

where

$$a_\mu = (E^\rho \delta_\mu^\sigma - \frac{1}{2}\epsilon_{\mu\nu}^{\rho\sigma} B^\nu)\sigma_{\rho\sigma} + \mathrm{i}F_{\mu\nu}\gamma^\nu. \qquad (4.28)$$

The phase factor in (4.27) is non-Abelian because (4.28) shows that a_μ is a non-Abelian gauge potential of the $SO(4,1)$ gauge group generated by $\sigma_{\rho\sigma}$ and $\mathrm{i}\gamma_\mu$. The phase shift due to this non-Abelian phase factor for interference around an infinitesimal closed curve around the area element $\Delta S^{\mu\nu}$ is given by

$$\Delta\theta = \frac{\mu}{2}\Sigma_{\rho\sigma}\,F_{\mu\nu}^{\rho\sigma}\,\Delta S^{\mu\nu}, \qquad (4.29)$$

where $F_{\mu\nu}^{\rho\sigma}$ is the Yang–Mills field defined by

$$F_{\mu\nu}^{\rho\sigma}\sigma_{\rho\sigma} = \partial_\mu a_\nu - \partial_\nu a_\mu + \mathrm{i}\mu[a_\mu, a_\nu], \qquad (4.30)$$

and $\Sigma_{\rho\sigma} = \bar{\psi}\sigma_{\rho\sigma}\psi = -\Sigma_{\rho\sigma}$. In the non-relativistic limit the gauge group reduces to $SU(2)$. In this sense the phase shift (4.29) is richer than the AB phase shift. It may be possible to verify such non-Abelian 'gauge field' effects experimentally, which has not been possible to do so far with weak and strong interactions because of their short-range character. The effect can also be generalized to the case when a gravitational field is present [101].

Again, the fact that the AAC effect is necessary for the complementarity of path distinguishability and interference has been demonstrated by Ramsey [102].

Goldhaber [103] has given a beautiful summary of the differences between the AB and AAC effects. They are:

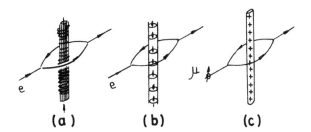

Fig. 4.9. Duality between the AB and AAC topologies (after Ref. [105]).

(1) For the AB effect the flux tube should be endless, but may be curved into a toroid, whereas for the AAC effect the line charge must be straight and parallel to the magnetic moment.

(2) For the AB effect the electromagnetic field strength vanishes in the region R accessible to the scattered particles, but for the AAC effect the electric field does not vanish, only the force on the particle vanishes for the correct alignment.

(3) There is an additional degree of freedom in AAC, namely the spin of the particle, that is analogous to the Yang–Mills potential of a particle carrying isospin but no spin, making such potentials observable in principle.

The duality between the AB topology and the AAC topology is illustrated in Fig. 4.9.

These differences make the AAC effect intrinsically richer in possibilities than the AB effect. For example, the Hamiltonian (4.19) shows that the simple and 'naive' force expression $\vec{\nabla}(\vec{\mu} \cdot \vec{B})$ should be replaced by the expression

$$\vec{F} = \vec{\mu} \cdot \vec{\nabla} \times \vec{B} + \vec{E} \times d\vec{\mu}/dt. \qquad (4.31)$$

This is necessary for the force-free, velocity-dependent interaction and follows from the conservation of total momentum, including the electromagnetic contribution $\vec{E} \times \vec{\mu}$. If one uses the naive expression $\vec{\nabla}\vec{\mu} \cdot \vec{B}$ for the force, the phase shift turns out to be much smaller than the AAC prediction from (4.31) [104]. Thus, an experimental confirmation of the AAC effect would more than just confirm the AB effect – it would give a subtle quantum check of the principle of total momentum conservation.

The AAC effect has, in fact, been confirmed experimentally in neutron interference experiments [105]. An unpolarized neutron beam of wavelength $\lambda = 1.477$ Å was passed through a perfect crystal interferometer containing a 30 kV/mm vacuum electrode system (Fig. 4.10). Since the AAC phase shift is topological and depends only on the linear charge

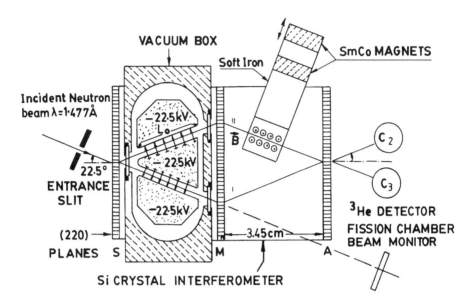

Fig. 4.10. Schematic diagram of the experiment to observe the AAC effect (after Ref. [105]).

density enclosed by the beam paths but not on any details of their geometry relative to the line charge, a charged prism-shaped electrode system was placed between the splitter (S) and the mirror (M) plates of the interferometer. Thus, the neutron wave packet in path II passed through a region of electric field $\vec{E}$, and then through a vertical magnetic bias field $\vec{B}$ which could be adjusted to obtain first-order sensitivity. The neutron wave packet in path I passed on the opposite side of the centre electrode whose polarity was periodically reversed. The interferometer, along with the entrance slit, the vacuum electrode cell and the bias magnet could be tilted about the incident beam direction so that the two neutron wave packets were at different heights and experienced different gravitational potentials, inducing a relative (spin-independent) gravitational phase shift $\Delta\phi_G$. This phase shift was adjusted to cancel the experimental offset phase shift $\Delta\phi_0$, giving maximum sensitivity. This cancellation is usually done by inserting an aluminium plate between two of the crystal slabs of the interferometer, but this was not feasible because of space limitations within the interferometer.

The theoretically predicted phase shift is 1.50 mrad for the geometry and the conditions of the experiment. The observed phase shift was 2.19 ± 0.52 mrad. Thus

$$(\Delta\phi_{AAC})_{exp}/(\Delta\phi_{AAC})_{theor} = 1.46 \pm 0.35. \qquad (4.32)$$

The accuracy of the experiment was limited by the available neutron intensity, running time and long-term apparatus stability. The errors are strictly statistical, leaving systematic errors for further identification and assessment. Significant improvements are possible with much larger neutron interferometers or atomic beam interferometers for which the magnetic moment is ~ 2000 times larger than for neutrons.

4.5   The geometric phase

It has now become clear that the AB and AAC effects are special cases of the 'geometric phase' that is generated in all cyclic evolutions because of the geometry of the projective space of rays of the Hilbert space $\mathscr{H}$ [106]. Let $|\psi\rangle$ be a normalized state in $\mathscr{H}$. Then for all cyclic evolutions one has

$$|\psi(\tau)\rangle = e^{i\phi}|\psi(0)\rangle \tag{4.33}$$

where ϕ is a real number. Let the projection map $\Pi : \mathscr{H} \to \mathscr{P}$ where $\mathscr{P}$ is the projective Hilbert space of rays of $\mathscr{H}$, be defined by $\Pi(|\psi\rangle) = \{|\psi'\rangle : |\psi'\rangle = c|\psi\rangle\}$, where c is a complex number. A ray is an equivalence class of states differing only in phase. This means that all state vectors in $\mathscr{H}$ that are related by a phase are mapped by Π on to the same ray in $\mathscr{P}$. Since $|\psi(t)\rangle$ satisfies the Schrödinger equation

$$i\hbar \frac{d}{dt}|\psi(t)\rangle = H|\psi(t)\rangle, \tag{4.34}$$

it defines a curve $C : [0, \tau] \to \mathscr{H}$ such that $\hat{C} \equiv \Pi(C)$ is a *closed* curve in $\mathscr{P}$. Conversely, given any such curve C, one can define a Hamiltonian function $H(t)$ so that (4.34) is satisfied for the corresponding normalized state $|\psi(t)\rangle$. Now, define

$$|\tilde{\psi}(t)\rangle = e^{-if(t)}|\psi(t)\rangle \tag{4.35}$$

such that $f(\tau) - f(0) = \phi$. Then $|\tilde{\psi}(\tau)\rangle = |\tilde{\psi}(0)\rangle$, and so it follows from (4.34) that

$$\frac{1}{\hbar}\langle\psi(t)|H|\psi(t)\rangle = \langle\tilde{\psi}(t)|i\frac{d}{dt}|\tilde{\psi}(t)\rangle - \frac{df}{dt}. \tag{4.36}$$

Hence, if one removes the dynamical part of the phase by defining

$$\beta = \phi + \frac{1}{\hbar}\int_0^\tau \langle\psi(t)|H|\psi(t)\rangle\,dt, \tag{4.37}$$

it follows from (4.36) that

$$\beta = \int_0^\tau \langle\tilde{\psi}|i\left(\frac{d}{dt}|\tilde{\psi}\rangle\right)dt. \tag{4.38}$$

Clearly, the same $|\tilde{\psi}(t)\rangle$ can be chosen for every curve C for which $\Pi(C) = \hat{C}$ by an appropriate choice of $f(t)$. Hence, β (4.37) is independent of ϕ and H for a given closed curve $\hat{C}$. It is therefore universal in the sense that it is the same for all possible motions along curves C in the Hilbert space $\mathscr{H}$ which project to a given closed curve $\hat{C}$ in $\mathscr{P}$. It also follows from (4.38) that β is independent of the parameter t of $\hat{C}$, and is uniquely defined up to $2\pi n$ ($n =$ integer).

One may regard $\mathscr{H}$ as a fibre bundle over $\mathscr{P}$. The bundle has a natural connection which permits a comparison of the phases of states on two neighbouring rays. When the dynamical phase factor is removed, the evolution of the system is a parallel transport of the phase of the system according to this natural connection. The phase β is then a consequence of the curvature of this connection, and is kinematic in origin and geometrical in nature [107].

Originally, Berry [108] had considered only adiabatic cyclic evolution. The geometric phase β is a generalization of the Berry phase for non-adiabatic evolutions as well. It is also possible to obtain a gauge-invariant generalization of β [109]. Furthermore, it has been shown that the evolution need neither be cyclic nor unitary and may be interrupted by quantum measurements [110].

The subject has developed rapidly with applications in many branches in physics and chemistry. A fuller account of the developments will be found in *Geometric Phases in Physics* [111].

5

Macroscopic quantum coherence

5.1 The measurement problem

Schrödinger's cat paradox is perhaps the most dramatic example that highlights the acuteness of the conceptual difficulties inherent in extending the superposition principle to macroscopic systems. We have already seen in Chapter 2, Section 2.3.1, how Schrödinger dramatized the transfer of the quantum mechanical superposition of microscopic states to a macroscopic object like a cat. The paradox has its roots in the quantum theory of measurement. Let us consider a 'pure state'

$$| \psi \rangle = \sum_i c_i | \phi_i \rangle, \quad \sum_i |c_i|^2 = 1. \tag{5.1}$$

If one measures a physical observable A on this state, one would get the result a_i with probability $|c_i|^2$, and once the measurement is complete, the state is forced into the eigenstate $| \phi_i \rangle$. This does not, however, imply that the system is actually a statistical ensemble of these states $| \phi_i \rangle$ with probability $|c_i|^2$ in the sense of classical probability theory, and that the measurement simply removes the ignorance. The simplest way to see this is to compare the density matrix of the state (5.1),

$$
\begin{aligned}
(\rho)_{ij} &= (| \psi \rangle \langle \psi |)_{ij} \\
&= c_i^* c_j
\end{aligned}
\tag{5.2}
$$

which is non-diagonal with that for a 'mixture',

$$(\hat{\rho})_{ij} = |c_i|^2 \delta_{ij}, \tag{5.3}$$

which is diagonal. Although (5.2) and (5.3) give identical results for the probabilities of obtaining the various eigenvalues a_i of an observable A, they predict quite different results for observables that do not commute

89

with A. For example, consider the spin state

$$| \chi \rangle = \frac{1}{\sqrt{2}} \left[| \frac{1}{2}, \frac{1}{2} \rangle + | \frac{1}{2}, -\frac{1}{2} \rangle \right]. \tag{5.4}$$

It is an eigenstate of the spin operator S_x with eigenvalue $+\frac{1}{2}$. Therefore, a measurement of S_x on this state is certain to give the result $+\frac{1}{2}$. However, (5.4) is not an eigenstate of the operator S_z which does not commute with S_x. If one measures S_z on this state, one would obtain the eigenvalues $\pm\frac{1}{2}$ with equal probability. This shows that the state (5.4) does not *have* a definite value of any observable we may choose to measure on it. On the other hand, if a system is described by the diagonal density matrix $\frac{1}{2}\delta_{ij}$, it *can* be interpreted as a statistical mixture of states *having* the values $\pm\frac{1}{2}$ of spin with equal probabilities. According to standard measurement theory a measurement of S_z quenches the off-diagonal interference terms and reduces the density operator of a pure state like (5.4) to the diagonal form $\frac{1}{2}\delta_{ij}$. The absence of interference between the various possibilities then makes it possible to interpret the diagonal elements as classical probabilities. This process of decoherence or loss of coherence of a pure state on measurement, or reduction of the state vector as it is often called, cannot result from Schrödinger evolution which is unitary, causal and reversible. It is described by a nonlocal, non-unitary, instantaneous, non-causal and irreversible process

$$\rho \rightarrow \hat{\rho} = \sum_i \Pi_i \rho \, \Pi_i$$

$$\Pi_i = | \phi_i \rangle \langle \phi_i |, \tag{5.5}$$

which is the least understood aspect of quantum mechanics. It lies at the origin of the many interpretations of quantum mechanics.

5.2   Interpretations

5.2.1   Bohr's interpretation

The preceding discussion fits in with the standard Copenhagen dictum that one must not think of microsystems as *possessing* properties in their own right irrespective of the experimental condition which is determined by some *classical* measuring apparatus. For example, if one wishes to measure S_z on the system (5.4), the classical apparatus required for doing so *ipso facto* precludes the simultaneous measurement of S_x on the same system. One may therefore, if one so desires, describe the system by the reduced density matrix $\hat{\rho} = \frac{1}{2}\delta_{ij}$ (which describes a 'mixture') even before the measurement is made, without any fear of being contradicted by experiment. What if one had opted to measure S_x instead? The usual answer, based on the complementarity principle, is that since the experimental conditions in the two cases are different and *mutually exclusive*, it is illogical to assume that the description of the microsystem remains the same. In other words, *at the microscopic level, the language of quantum mechanics is not that of realism.*

At the macroscopic level (the level of billiard balls and counters), on the other hand, Bohr and his Copenhagen school assert that objects *must be described by the language of classical physics, the language of realism.* As Bohr often emphasized, unless the preparation and measuring devices themselves are described by this language of realism and behave causally, we are unable at all to communicate our experimental procedures and results to one another *unambiguously*. It is an indisputable fact that the familiar objects of everyday experience do *possess* definite properties irrespective of specific experimental setups to measure them. No exception to this general feature of the macroscopic domain has ever been observed so far – the superposition principle appears to be violated by everyday classical reality. This is despite the fact that the actual values of the properties are ultimately determined by effects at the microscopic level which are intrinsically quantum mechanical in nature. For example, the fact that the Debye law of the specific heat of solids at low temperatures is a consequence of the quantum mechanical nature of the electrons in the solids, does not necessarily imply that superpositions of states of solids with widely different energies can occur. Similarly, although superconductivity can only be explained in terms of the quantum mechanical behaviour of electrons in low temperature solids forming Cooper pairs, it is perfectly adequate to regard the superconducting wave function of the centre of mass of Cooper pairs as a classical field whose value is well defined at each space–time point. This sharp *qualitative* distinction between microsystems (for which a realistic description is *forbidden*) and the classical measuring apparatus (for which a realistic description is *fundamental* (non-derivable)), lies at the heart of the Copenhagen interpretation. But, as Bohr emphasized, the border between the quantum and the classical must be left mobile, so that even the human nervous system, the ultimate apparatus, can also be measured and analyzed as a quantum system, provided a suitable classical device is available to do this. According to Bohr, *the notion of measurement as an operation cannot be analyzed within the framework of quantum mechanics itself.*

5.2.2 *von Neumann's interpretation*

If one is not willing to be content with the dichotomy between the quantum and the classical, one must be prepared to consider the possibility that, in principle, quantum mechanics should apply to all physical objects including state preparation and measuring devices and the human nervous system. According to this alternative point of view, first taken seriously by von Neumann [112], quantum mechanics is *in principle a complete and universal theory of the entire physical world*. Consequently, von Neumann had to contend with the following situation. Suppose that a microsystem is initially in a pure state such as (5.1) and the apparatus is in a definite eigenstate $|X_0\rangle$ of some operator. If the time-dependent Schrödinger equation correctly describes the interaction between the microsystem and the measuring apparatus, then the linearity of the time evolution operator would inevitably imply an evolution of the form

$$\left(\sum_i c_i |\phi_i\rangle\right) |X_0\rangle \rightarrow \sum_i c_i |\phi_i\rangle |X_i\rangle. \tag{5.6}$$

This shows that under certain conditions the 'pure' character of the microsystem can get transmitted to the measuring apparatus which is then described by a linear superposition, each state of the apparatus being correlated to a particular state of the microsystem. In other words, *the apparatus ends up in a Schrödinger cat state*. That certainly conflicts with our commonsense idea of a macroscopic object. Moreover, there is no uniqueness about the apparatus basis $|X_i\rangle$; there is nothing in quantum mechanics to prevent the appearance of some other complete set of basis states $|X_i'\rangle$ for the apparatus in (5.6). What then is responsible for the choice of a preferred basis in a measurement?

The orthodox Copenhagen analysis would proceed as follows. Choose an interaction between the microsystem and the measuring apparatus (treated quantum mechanically in this case) such that the final states of the apparatus are not only *orthogonal* ($\langle X_i | X_j \rangle = \delta_{ij}$), they are also *macroscopically distinguishable* as judged by a classical device such as a pair of human eyes. An example is the clearly resolved pair of spots on a screen at a suitable distance from the magnets in a Stern–Gerlach set up to measure the spin of a neutral atom. Such a condition is *necessary* for a good measurement to be at all possible. Then one can show that the various possible eigenstates $|\phi_i\rangle$ of the microsystem to which the apparatus states $|X_i\rangle$ (preferred by the classical observer) are correlated, decohere to a high degree of approximation. They can also be read off from the corresponding states of the apparatus without any ambiguity.

The von Neumann approach, however, proceeds differently. Since a unitary and causal Schrödinger evolution (which von Neumann calls 'Process 2') is assumed to provide a complete and universal description of all physical processes, it can never reduce an isolated pure state to a mixture. He therefore postulated an *additional* non-unitary, non-causal process ('Process 1') that an *observation* of the apparatus brings about by reducing the density matrix ρ of the combined system (I(microsystem) + II(measuring apparatus)) to a diagonal form $\hat{\rho}$ (5.5). This is the famous projection or collapse postulate that renders a classical probability interpretation of the various measurement outcomes possible. 'Process 1' is only applicable to the instantaneous and irreversible measurement interaction between (I + II) and III(the observer). Thus, 'the measurement and the related process of the subjective perception is a new entity relative to the physical environment and is not reducible to the latter'. This dual description in terms of two independent processes is necessitated by the formalism of a quantum mechanical description of nature, and replaces the duality between the quantum and the classical in the physical world that is central to Bohr's's interpretation.

The question that arises is: where does one draw the boundary between the apparatus and the observer? One obviously needs a second apparatus to make a measurement on the first, a third apparatus to make a measurement on the second, and so on. Where does this sequence stop? von Neumann showed that the apparatus can be extended to include the entire physical world (including the sense organs of the observer, his nervous system, brain and so forth, but *excluding* his 'abstract ego' which becomes conscious of the result) without making any difference to the result for I. In other words, as in Bohr's interpretation, the boundary between the system and the apparatus or between the apparatus and the sense organs or between the sense organs and the nervous system and so on and so forth, is mobile. In von Neumann's words, 'The fact that the boundary can be pushed arbitrarily deeply into the interior of the body of the actual observer is the content of the principle of psycho-physical parallelism', a principle which he regarded as a 'fundamental requirement of the scientific viewpoint'. It states that 'it must be possible so to describe the extra-physical process of the subjective perception as if it were in reality in the physical world – i.e. to assign to its parts equivalent physical processes in the objective physical environment, in ordinary space'. The difference with Bohr is that the ultimate observer is not a classical device in the physical world but the 'abstract ego' of a human observer which is independent of the physical world.

5.2.3 The many worlds interpretation

Is there a way out without invoking a subjective experience external to the physical world? In other words, is it possible to generalize the framework of quantum mechanics so as to apply it to the universe as a whole, i.e., to cosmology? It was Everett [113] who first suggested how to do this in his famous 'relative state' formulation in which the observer (be it an inanimate instrument or the consciousness of a human being) is included as part of the physical world. The wave function then describes the entire universe (or any closed system) that evolves by 'Process 2' alone. 'Process 1' together with the standard probabilistic interpretation are *derived* from this 'metatheory' by simply allowing the physical system representing the observer (which is in a superposition of states such as (5.6) that are correlated with 'relative' states of the system) to 'branch' into a number of different states, each branch representing a different outcome of the measurement. 'All branches exist simultaneously in the superposition after any given sequence of observations.' However, because these branches are orthogonal, 'there is a total lack of effect of one branch on another' so that 'no observer will ever be aware of any "splitting" process' of the world into 'many worlds', and consequently there is no need to suppose that all but one branch are somehow destroyed.

Though very elegant as a 'metatheory' that justifies the standard interpretation, it fails the test of Occam's razor – there is a conceptual proliferation of unrelated worlds at every observation. It is also not clear how it resolves the difference between the quantum system and the measuring apparatus inherent in the Copenhagen interpretation. Consequently, it is not at all clear when and how the branching takes place.

5.2.4 *Emergence of the classical world*

It is possible, however, to understand the emergence of a quasi-classical world *as an approximation*, without invoking either a conscious observer outside the physical world or the many worlds interpretation. Consider two interacting systems A and B described by the pure state

$$|\psi\rangle = \sum_i c_i |\phi_i(A)\rangle |\chi_i(B)\rangle. \tag{5.7}$$

There is a simple theorem that states that such a state cannot be distinguished from a 'mixture' of the states $|\phi_i(A)\rangle |\chi_i(B)\rangle$ with probabilities $|c_i|^2$ by any measurement carried out on one of the systems only. The proof is simple. Take the density matrix ρ of $|\psi\rangle$ and trace it over the states of A (B), remembering that they are orthogonal – one obtains the reduced density matrix $\hat{\rho}$ (5.5) of B (A). An application of this theorem to (5.6) shows that the right-hand side is equivalent to a 'mixture' as regards any subsequent measurement carried out on the microsystem alone.

It is, however, possible to make such a distinction provided one can measure correlations of the form $\langle \hat{O}(A)\hat{P}(B)\rangle$ where $\hat{O}$ operates only on A and $\hat{P}$ only on B, and provided both $\hat{O}$ and $\hat{P}$ are non-diagonal in the representation labelled by ϕ_i and χ_i. Is it possible actually to demonstrate such correlations between the macroscopically different states $|X_i\rangle$ of a macroscopic apparatus? Much effort has gone into showing that it is not, essentially because a macroscopic object contains a very, very large number N' of interacting atoms, electrons and so on, and is extremely sensitive to its *environment*. Consider two macroscopically distinguishable states of the apparatus. They are distinguishable because a certain number $N < N'$ of the constituents must be in different states in them. To distinguish a superposition of such states from a mixture, one must measure at least the expectation value of an N-particle operator, which is totally out of the question since N is inevitably very large because of the unavoidable interactions of a macroscopic body with its environment. Daneri, Loinger and Prosperi [114] have attempted to show that this is even necessary for a measurement to be at all possible by a thermodynamically stable state of the apparatus. This approach is in the spirit of Bohr, the only difference being that there is an attempt here at analyzing the emergence of the classical reality from an underlying quantum mechanical substratum, something that Bohr considered impossible.

Other attempts have also been made to explain the emergence of the classical world by considering the unavoidable interactions of macroscopic objects with their environment [115], [116], [117], [118]. The essential point is the idea of 'decoherence' (the approximate diagonalization of the density matrix) or the loss of coherence which results from tracing the pure density matrix of the total system (object plus environment) over the degrees of freedom of the environment. Although this is a useful approach in analyzing the mesoscopic domain, it does not really help to bring the measurement process within quantum mechanics itself as claimed by Zurek [118] for the following reasons. First, the concepts of 'system', 'apparatus' and 'environment' already imply an arbitrary division of the world (prior to the application of the Schrödinger equation) that is not precisely defined if quantum mechanics is universally valid. Second, since a unitary evolution maps only pure states into pure states, the transition from a pure to a mixed state by Schrödinger evolution can only be an approximation, albeit an excellent one in practice in some cases. For critical appraisals of this approach see Refs. [119], [120], [121].

5.2.5 *The many histories interpretation*

Gell-Mann and Hartle [122], [123], [124] have developed an interesting programme to understand the emergence of the quasi-classical domain of familiar experience and the process of measurement from quantum cosmology. Theirs essentially is an attempt at 'extension, clarification and completion of the Everett interpretation' which, as we have seen, was incomplete. 'It did not adequately explain the origin of the classical domain or the meaning of the 'branching' that replaced the notion of measurement. It was a theory of 'many worlds' ..., but it did not sufficiently explain

how these were defined or how these arose' [124]. Gell-Mann and Hartle
preferred to call their own extension a 'many histories' interpretation.

Griffiths [125] was the first to introduce the notion of 'alternative
histories' of a quantum system by means of alternative sequences of
time-ordered projection operators corresponding to observables describing specific information. Let $P(t)$ be the projection operator in the
Heisenberg representation corresponding to some observable $\mathcal{O}(t)$. Then
$(P_1^k(t), P_2^k(t), \ldots)$ is a set of such projection operators, the super-
script k labelling the set, the subscript α the particular alternative and t
the time. An *history* is a particular sequence of alternatives $[P_\alpha] = (P_{\alpha_1}^1(t_1), P_{\alpha_2}^2(t_2), \ldots, P_{\alpha_n}^n(t_n))$. Griffiths showed that only histories that
satisfied a certain necessary criterion (the vanishing of the real part of
every interference term between sequences) were 'consistent', and could
be assigned probabilities that were additive like classical probabilities.
Omnès [126], [127], built his 'logical interpretation' of quantum mechanics
on this idea by showing that the rules of ordinary logic could be recovered when making statements about properties that satisfied the Griffiths
criterion.

Gell-Mann and Hartle introduce three new elements in their approach:
(a) the notion of sets of alternative coarse-grained histories of a quantum
system, (b) the 'decoherence' of the coarse-grained histories in a set, and
(c) their approximate determinism near the classical limit. Coarse-graining
can be done in three ways: (1) by specifying the observables at some times
only, not at all times; (2) by specifying some but not all observables of
a complete set at one time; and (3) by specifying for these observables
not precise values but only ranges of values. Only suitably coarse-grained
histories decohere, not completely fine-grained histories in the coordinate
basis. The rule for determining whether probabilities can be assigned to a
given set of alternative histories, and what these probabilities are, is given
by an important theoretical construct called the decoherence functional
$D[(\text{history})', (\text{history})]$:

$$D([P_{\alpha'}], [P_\alpha]) = \text{Tr}\left[P_{\alpha'_n}^n(t_n) \ldots P_{\alpha'_1}^1(t_1) \rho P_{\alpha_1}^1(t_1) \ldots P_{\alpha_n}^n(t_n) \right], \quad (5.8)$$

where ρ is the original density matrix of the universe. A set of coarse-
grained histories decohere when the off-diagonal elements of D are sufficiently small:

$$D([P_{\alpha'}], [P_\alpha]) \approx 0, \quad \text{for any } \alpha'_k \neq \alpha_k. \quad (5.9)$$

This is a generalization of the condition, for example, of the absence of
interference in the double-slit experiment. It is a sufficient condition and is
stronger than the Griffiths criterion for consistent histories. Probabilities
can be assigned to the individual members of a set of alternative histories
to the extent that it decoheres.

Given the initial condition of the universe and the action function of the elementary particles, it is possible in principle to predict the probabilities for any set of alternative decohering coarse-grained histories of the universe using this programme. However, as Gell-Mann and Hartle comment [124]:

> It is an important question whether all the quasiclassical domains are roughly equivalent or whether there are various essentially inequivalent ones. A measurement is a correlation with variables in a quasiclassical domain. An 'observer' (or information gathering and utilizing system) is a complex adaptive system that has evolved to exploit the relative predictability of a quasiclassical domain, or rather a set of such domains among which it cannot discriminate because of its own coarse graining. We suggest that resolution of many of the problems of interpretation presented by quantum mechanics is to be accomplished, not by further scrutiny of the subject as it applies to reproducible laboratory situations, but rather by an examination of alternative histories of the universe, stemming from its initial condition, and a study of the problem of quasiclassical domains.

5.2.6 The de Broglie–Bohm (dBB) interpretation

An interpretation of considerable current interest is the ontological and causal interpretation of de Broglie and Bohm [128], [129]. It differs from the interpretations discussed so far in regarding the wave function as an incomplete description of a quantum system and treating the position x of a particle as an additional (so-called hidden) variable to complete the description. It is similar to the many worlds and many histories interpretation in regarding the wave function (augmented by the position variable) as a description of a closed system. Unlike the other interpretations, it is a realist interpretation in which a particle *exists* all the time with all its properties, and the mysterious measurements are reduced to ordinary interactions which occur all the time in real processes during which the actual values of the properties evolve continuously into appropriate eigenvalues in accordance with 'Process 2' alone. It is constructed to give the same predictions as quantum mechanics in all situations where the latter makes unambiguous predictions. We will return to further discussions of this interpretation in Chapter 10 while discussing the problem of tunneling times.

5.2.7   *The Ghirardi–Rimini–Weber–Pearle (GRWP) model*

Of course, it is not a priori clear that the linear laws of quantum mechanics *are* universal and *must* apply to every physical object, however macroscopic and complex it might be. If they do not, the problem of distinguishing cat states from mixtures does not arise in the first place. It is quite conceivable that beyond a certain degree of complexity nonlinear terms begin to play an important role, rapidly reducing cat states to one of their branches [130]. We will have occasion to discuss this model in greater detail in Chapter 7.

Meanwhile, the most important question that must be answered is: *what experimental evidence is there or could there be that macroscopic objects actually can occur in pure states of the form (5.6)?* An empirical answer to this question is of vital importance for the quantum theory of measurement. In looking for an answer one has to bear in mind the essential difference between *macroscopic quantum tunneling* (MQT) and *macroscopic quantum coherence* (MQC) which we consider in the next two sections.

5.3   Macroscopic quantum tunneling

The first thing to look for would be to see if quantum mechanics works at the macroscopic level at all, without looking for coherence effects. One possibility is to look for a macroscopic analogue of the tunneling of an α-particle out of a heavy nucleus [131]. Consider a potential $V(q)$ (Fig. 5.1) which has a local metastable minimum separated from an unbounded region of considerably lower potential energy ($V(q) < 0$ for all $q > 0$) by a classically impenetrable barrier of height U_0 and width Δq. Suppose that the variable q of a system is known to have a value within the metastable well at some initial time. The probability that the system will be found within the well at later times which are neither too short nor too small decreases exponentially with a decay constant Γ if the system behaves quantum mechanically.

There are three features that make such tunneling differ from α-decay. First, if values of q differing by Δq can be regarded as 'appreciably different' by some reasonable criterion, the tunneling process would constitute evidence of tunneling between 'macroscopically different' states, and therefore provide circumstantial evidence that quantum mechanics works at the macroscopic level. This is the main point of interest. Second, as we have seen, a macroscopic variable is inevitably coupled to its environment, and this feature must be taken into account in calculating the decay rate [132]. Third, because the characteristic frequency of the macroscopic variable ($\omega \sim M^{-1/2}$ where M is the associated inertia) is low, it is easy for the

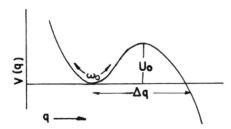

Fig. 5.1. Macroscopic quantum tunneling (after Ref. [133]).

system to escape from the well by thermal fluctuations rather than by quantum tunneling. Therefore, the temperature dependence of the decay rate must also be calculated.

Let us consider the case of an r.f. Superconducting Quantum Interference Device (SQUID) ring. The trapped flux Φ plays the role of the macroscopic variable q. If one ignores for the time being its coupling to the environment, its Lagrangian can be written as

$$L = \frac{1}{2C} p_\Phi^2 - U(\Phi)$$

$$U(\Phi) = \frac{(\Phi - \Phi_x)^2}{2L} - \frac{1}{2\pi} J_c \Phi_0 \cos\left(2\pi \frac{\Phi}{\Phi_0}\right), \qquad (5.10)$$

where Φ_x is the externally applied flux which is treated as a c-number, J_c is the critical current of the Josephson junction, L is the self-inductance of the ring, and Φ_0 is the flux quantum $h/2e$; the capacitance C of the junction plays the role of a particle 'mass' [131]. If the quantity $\beta_L \equiv 2\pi L J_c / \Phi_0 > 1$, the potential (5.7) has more than one metastable minimum. Although the potential in the region 'outside the barrier' is never infinite in this case, for $\beta_L \gg 1$ it is so large in comparison with the potential in the 'inside region' that it should be good approximation to regard it as infinite [132]. The case β_L close to one actually corresponds to the MQC setup which we will discuss in Section 5.4.

Since most experiments have actually been done on current-biased junctions, it is convenient to take $\beta_L \to \infty$ in which limit the SQUID ring is identical to a current-biased junction with the replacements $\Phi \to (\Phi_0/2\pi)\Delta\phi$, $\Delta\phi$ being the phase difference between the condensate wave function on the two sides of the junction, and $\Phi_x \to L J_x$ where J_x is the externally imposed current [133]. One is usually interested in the case where J_x is close to the critical current J_c so that the height of the barrier is small compared to $J_c \Phi_0 / 2\pi$. In this limit one can show [133] that the barrier height U_0 and the small oscillation frequency ω_0 are given by

$$U_0 = \frac{2\sqrt{2}}{3\pi} \Phi_0 J_c (\delta J/J_c)^{3/2}, \tag{5.11}$$

$$\omega_0 = (2\pi J_c/C\Phi_0)^{1/2} (2\delta J/J_c)^{1/4}, \tag{5.12}$$

where $\delta J \equiv J_c - J_x$. Furthermore, in this limit

$$U(q) = \frac{3}{2} U_0 \left[\left(\frac{q}{\Delta q}\right)^2 - \left(\frac{q}{\Delta q}\right)^3 \right], \tag{5.13}$$

where we have put $\Delta\phi \equiv q$ and written the width of the barrier as Δq. The quantum tunneling rate at zero temperature can be calculated from this potential for an isolated system by the WKB technique.

When account is taken of the coupling with the environment, the decay rate is given by

$$\Gamma = A \exp(-B/\hbar), \tag{5.14}$$

where the constants A and B can be calculated in various approximations [132]. This shows that dissipation of energy leads to a suppression of the probability of tunneling, as expected. The temperature dependence of the decay rate can also be estimated at various ranges of temperature for both strong and weak damping [133].

Several experiments have been performed on macroscopic quantum tunneling in current-biased junctions and in SQUIDs [134], [135], [136]. The order of magnitude of the measured tunneling rates and their dependence on temperature and resistance are in qualitative agreement with the theoretical predictions, confirming the existence of MQT and paving the way for the more interesting and challenging task of looking for MQC.

5.4   Macroscopic quantum coherence

One of the typical consequences of the superposition principle in quantum mechanics is oscillatory behaviour, such as that observed in ammonia molecules, neutral kaon systems and other microsystems, between two distinct but nearly degenerate states. If the superposition principle is applicable to macroscopic systems, one would expect such systems also to show analogous behaviour which has been termed 'Macroscopic Quantum Coherence' (MQC). In order to look for such behaviour one needs a system with a macroscopic variable q in a symmetric double-well potential $V(q)$ which has two degenerate or nearly degenerate minima (say at $\pm q_0/2$) and a barrier height U_0 (Fig. 5.2). Let the kinetic energy be $\frac{1}{2} M \dot{q}^2$, and let the quantity $[M^{-1}(\partial^2 V/\partial q^2)]^{1/2}$ be denoted by ω_0. Let us assume that (a) the zero point energy $\hbar\omega_0/2$ in each of the two wells is smaller than

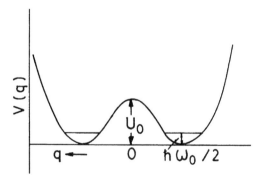

Fig. 5.2. Macroscopic quantum coherence (after Ref. [133]).

the barrier height U_0 so that the WKB approximation can be applied to the motion under the barrier, and that (b) $k_B T$ is very small compared to $\hbar\omega_0$. These conditions guarantee that as well as all but the lowest energy states in the two wells, the classical Kramers process of transition above the barrier are negligible.

If $U_0 \to \infty$, the system would be either in the left well or the right well. Let these states be denoted by ψ_L and ψ_R. They are degenerate with energy $\frac{1}{2}\hbar\omega_0$ relative to the bottom of the wave. For any finite value of $U_0/\hbar\omega_0$ there is a finite amplitude for tunneling denoted by $\frac{1}{2}\Delta$ given by the formula

$$\frac{1}{2}\Delta = const. \omega_0 \exp\left(-\int[2M\,V(q)]^{1/2}\,dq/\hbar\right)$$

$$= const. \omega_0 \exp\left(-const. U_0/\hbar\omega_0\right), \qquad (5.15)$$

which removes the degeneracy, so that one gets a doublet of states ψ_+ and ψ_- with energies E_+ and E_- given by

$$\psi_+ = \frac{1}{\sqrt{2}}(\psi_L + \psi_R), \quad E_+ = \frac{1}{2}\hbar\omega_0 - \frac{1}{2}\Delta, \qquad (5.16)$$

$$\psi_- = \frac{1}{\sqrt{2}}(\psi_L - \psi_R), \quad E_- = \frac{1}{2}\hbar\omega_0 + \frac{1}{2}\Delta. \qquad (5.17)$$

Since ψ_L is a linear combination of ψ_+ and ψ_- with slightly different energies, the system will oscillate between the two wells with a frequency $2\pi/\Delta$. Then, provided it is in the left well at $t = 0$, it will definitely be found in the left well at all times that are integral multiples of $2\pi/\Delta$, and in the right well at times that are half-odd-integral multiples of $2\pi/\Delta$. At all other times there will be a finite probability of finding it in either well. Thus, the probability of finding it in the left well minus the probability of finding it in the right well is given by

$$P(t) = \cos(\Delta t). \qquad (5.18)$$

The probability of actually finding the system anywhere between the two wells is practically zero ($\sim \exp(-U_0/\hbar\omega_0)$).

Suppose one arranges an experiment in which the system is trapped in a symmetric potential with two minima at $\pm q_0/2$ that are macroscopically distinct by some agreed criterion, and the system is in the left well at time $t = 0$. If the system is allowed to evolve freely for some time t and then a measurement is made on it to determine its state, it will be found to be in either of the two wells, the probability of finding it in any other state being negligible. Let a value $\pm$ be assigned every time q is found to be $\pm q_0/2$. If the experiment is repeated many times and t is varied, one would be able to obtain an experimental form of $P(t)$. What would one learn from that?

Notice that at time $t = \pi/(2\Delta)$, $P(t) = 0$ according to (5.18) which is a quantum mechanical prediction. This means there is equal probability of finding the system in either of the two wells at $t = \pi/(2\Delta)$. The same conclusion could be drawn quantum mechanically if the system happened to be actually a statistical mixture of the two states ψ_L and ψ_R at time $t = \pi/(2\Delta)$, as one would expect for a macroscopic system. How would one distinguish between these two possibilities? The answer is simple. If the system *were* a statistical mixture at some time, its Schrödinger evolution would guarantee that it would remain so for all subsequent times, so that quantum mechanics would predict $P(t) = 0$ for all subsequent times. On the other hand, if the system *were* in a linear superposition of the states ψ_L and ψ_R at $t = \pi/(2\Delta)$, $P(t)$ would behave as predicted by (5.18) at subsequent times. Therefore, if the experiment described above shows that $P(t)$ has the form (5.18), one would be able to *exclude* the possibility that the system was a statistical mixture of ψ_L and ψ_R at $t = \pi/(2\Delta)$ and conclude that it was indeed in a linear superposition of two macroscopically distinct states. Unfortunately, no definite experimental result is yet available.

This is analogous to the situation with Furry's hypothesis concerning the Einstein–Podolsky–Rosen paradox which will be discussed later (Chapter 9). As we will see, the famous EPR paradox arose in the context of two distant and non-interacting particles that had interacted in the past and got entangled in a state which is a linear superposition of two states each of which is a product of the states of the individual particles. Furry [137] suggested that such pure states could be experimentally distinguished from statistical mixtures, and therefore the correct description of the correlated system (a linear superposition or a mixture) could be experimentally determined. Some experiments were performed which excluded the mixture hypothesis.

Notice that all these discussions regarding pure and mixed states and measurements on them are valid *within the framework of quantum mechanics*. It was Bell [210] who proved a much stronger result in 1964. He showed that under certain conditions, an experimental result that agreed with the prediction of quantum mechanics would necessarily exclude a whole class of theories of widely separated but correlated microsystems (the 'objective local theories') *without reference to any quantum mechanical considerations*. A number of experiments were subsequently done which, in the opinion of the majority of physicists, excluded the objective local theories.

A remarkable result similar to Bell's has been obtained by Leggett and Garg [138] which can distinguish between MQC and a whole class of 'macro-realistic' theories that do not make use of any quantum mechanical considerations. This class of theories makes two fundamental assumptions that one would consider as part of commonsense outside a physics laboratory:

(A1) *Macroscopic realism*: a macroscopic system with two or more macroscopically distinct states available to it will at all times *be in one* or the other of these states.

(A2) *Non-invasive measurability at the macroscopic level*: it is possible, in principle, to determine the state of the system with arbitrarily small perturbation on its subsequent dynamics.

The surprising result that Leggett and Garg found was that *the conjunction of (A1) and (A2) is incompatible with the quantum mechanical prediction (5.18) for MQC*. Consider a quantity $Q(t)$ which has the value $+1$ if the system is in the left well, and the value -1 if it is in the right well. According to assumption (A1), the system will actually *have* the value ± 1 *at all times* t. Suppose an experiment is started with such a system *in some definite state* at time $t = 0$ and *left unobserved for some time*. Then it is straightforward to establish the inequality

$$Q(t_1)Q(t_2) + Q(t_2)Q(t_3) + Q(t_3)Q(t_4) - Q(t_1)Q(t_4) \leq 2 \quad (5.19)$$

by exhausting all the sixteen possibilities for the values of $Q(t_i)$ measured at times t_i. If it were possible by some means to measure the expectation values of all the quantities on the left-hand side of (5.20) *on the same ensemble*, one could predict the following result for the expectation values $K_{ij} \equiv \langle Q(t_i)Q(t_j)\rangle$,

$$K_{12} + K_{23} + K_{34} - K_{14} \leq 2, \quad (5.20)$$

which is analoguous to Bell's inequality (see Section 9.2.2). On the other hand, the *quantum mechanical prediction* for an *arbitrary* ensemble which is left undisturbed between t_i and t_j is, as we have seen, simply $K_{ij} = P(t_j - t_i)$ where $P(t)$ is the function (5.18). It is simple to verify that for certain values of t_j (for example, $t_2 - t_1 = t_3 - t_2 = t_4 - t_3 = \pi\Delta/4$ so that $t_4 - t_1 = 3\pi\Delta/4$) this quantum mechanical result violates the inequality (5.20). The times t_i here play the role of the polarizer settings in Bell's argument.

It might appear that this is sufficient to establish that quantum mechanics can be experimentally distinguished from macro-realistic theories. A closer scrutiny, however, reveals that it is not so without a further assumption, namely (A2). The reason is that in the actual experiment of interest, it is not known beforehand whether the system satisfies (A1) or MQC. The quantum mechanical prediction for K_{ij} for a system satisfying MQC holds only if the system is left undisturbed between t_i and t_j. This will certainly not be true if all the $Q(t_i)$ are 'measured' in a single run *on the same ensemble*. Any such 'measurement' on *any* system would inevitably satisfy the inequality (5.20) which is derived on the assumption that the system *is* in a definite state at the times t_i. So, one would have to *start* the system off in a definite state at time t_1, leave it undisturbed for a while and measure $Q(t_2)$ at a later time t_2, then start the system off again in the state it was in at t_1 but this time measure $Q(t_3)$ at time t_3 ($t_3 > t_2$), and so on. Unfortunately the measurements on the first and second groups of runs will then be on *ensembles with different properties* because the measurement on the first ensemble at t_2 inevitably changes its properties thereafter whereas no measurement is made on the second ensemble until t_3. The inequality (5.20) can therefore no longer be derived. This is why assumption (A2) (which is analogous to the assumption of locality in Bell's theorem) is required, namely that *it is possible, in principle, to make a measurement on an ensemble satisfying (A1) without changing its properties*. With such an assumption the ensembles become identical and inequality (5.20) *can* be derived for macro-realistic theories. The conjunction of (A1) and (A2) is therefore necessary for incompatibility with MQC. This completes the proof for an ideal macroscopic system totally isolated from its environment.

As we have already seen, it is extremely difficult in practice to isolate a macroscopic system from its environment. This necessitates a certain modification of the above argument for realistic experiments. Of course, the inequality (5.20) remains valid for all macro-realistic theories because it is a purely algebraic result. Only the quantum mechanical prediction (5.18) requires some modification. One can show quite generally [139] from measurement theory that

$$P(t) \sim e^{-\gamma t} \qquad (5.21)$$

in the presence of substantial system–environment interaction. This shows that even the quantum mechanical prediction automatically satisfies the inequality (5.20). So, what is the way out?

Leggett and his collaborators [140] have carried out extensive calculations in the real-life SQUID ring context. They verified the general result (5.21) in most cases. They define a dimensionless parameter

$$\alpha \equiv \eta \, q_0^2 / 2\pi\hbar \qquad (5.22)$$

where η is a friction coefficient. It can be written as $(\Delta\Phi/\Phi_0)^2 (R_0/R)$ in the SQUID case where $\Delta\Phi$ is the difference in trapped flux between the two minima of $U(\Phi)$ and R_0 is the 'quantum unit of resistance', $h/4\,e^2 \, (\sim 6\,\text{k}\Omega)$. The other relevant parameter is the temperature T. They showed that (5.21) is valid almost everywhere in the parameter space except in a small corner defined by the conditions

$$\alpha \ll 1, \qquad \alpha \, k_B T / \Delta_r \ll 1 \qquad (5.23)$$

where Δ_r is the tunneling frequency renormalized downwards by a large factor because of the coupling with the environment. In this region the system shows weakly damped oscillations analogous to those predicted by (5.18). Thus by keeping α and T small enough it should be possible in an actual SQUID ring experiment to observe violations of (5.20) should they occur, and thus learn from Nature whether she is at heart macro-realistic or quantum mechanical.

The paper of Leggett and Garg [138] generated a controversy on whether the magnetic flux trapped in an r.f. SQUID can be measured non-invasively [141], [142], [143]. Tesche [144] then showed how to design a *gedanken* experiment in which the joint probability densities for the magnetic flux trapped in a SQUID can be non-invasively determined using variable threshold superconducting switches followed by superconducting magnetometers.

Subsequently, an important conceptual clarification was achieved by Hardy *et al.* [145] who succeeded in constructing a class of realistic models which can be made to mimic the predictions of quantum mechanics for two-state oscillations by violating non-invasive measurability. This showed clearly that *the incompatibility in question was not between quantum mechanics and realism* (A1) *but arises solely due to the additional assumption of non-invasive measurability* (A2), contrary to Leggett and Garg's view that 'assumption (A2) is . . . a natural corollary of (A1) (realism)' [142].

6

The quantum Zeno paradox

6.1 Introduction

Quantum theory predicts the striking and paradoxical result that when a system is continuously watched, it does not evolve! Although this effect was noticed much earlier by a number of people [146], [147], [148], [149], [150], it was first formally stated by Misra and Sudarshan [151] and given the appellation 'Zeno's paradox' because it evokes the famous paradox of Zeno denying the possibility of motion to a flying arrow. It is as startling as a pot of water on a heater that refuses to boil when continuously watched. This is why it is also called the 'watched pot effect'. The effect is the result of repeated, frequent measurements on the system, each measurement projecting the system back to its initial state. In other words, the wave function of the system must repeatedly collapse. It is also necessary that the time interval between successive measurements must be much shorter than the critical time of coherent evolution of the system, called the Zeno time [152]. For decays this Zeno time is the time of coherent evolution before the irreversible exponential decay sets in, and is governed by the reciprocal of the range of energies accessible to the decay products. For most decays this is extremely short and hard to detect [153]. In the case of non-exponential time evolution, the relevant Zeno time can be much longer.

The paradox is of a different nature from the EPR and Schrödinger paradoxes. The latter are paradoxes of interpretation, there being no doubt as to the prediction of quantum theory. But when the quantum Zeno paradox was first pointed out, there was doubt as to whether it was a curious mathematical result devoid of any physical significance, merely the artefact of assuming continuous observation (as the limiting case of repeated observations) to be operationally meaningful, or an unequivocal prediction of the theory. The experimental verification of the effect by Itano *et al.* [154] settled this issue in 1990.

There is a closely related effect called the 'watchdog effect' in which the time evolution of a system is affected by its continuous coupling to the environment. In such a case the Hamiltonian is modified, and the effect is not as mysterious and counterintuitive as the 'watched pot effect'. It is the 'reduction of the state vector' caused by measurement that has haunted the foundation of quantum mechanics like a ghost. In spite of the claim made by Itano *et al.* [154] that the 'watched pot' effect has been observed in an experiment, this ghost continues to haunt quantum mechanics.

6.2 Zeno's paradox in quantum theory

Let us consider the quantum theory of an unstable system which decays into certain states. The total space of all possible states, the Hilbert space of the system $\mathcal{H}$, must include these decay products. Let the (orthogonal) projection onto the subspace spanned by the undecayed states of the system be E. Let the system be prepared in a state described by the density matrix ρ at time $t = 0$. Then, as a result of the Schrödinger evolution of the system described by the unitary operator $U(t) \equiv \exp(-iHt)$, the probability that the system will be *found* (i.e., observed) undecayed *at the instant t* is given by

$$q(t) = \text{Tr}\left[\rho\, U^*(t)\, E\, U(t)\right]. \tag{6.1}$$

Correspondingly, the probability that *at the instant t* the system will be *found* to have decayed must be given by

$$p(t) = 1 - q(t) = \text{Tr}\left[\rho\, U^*(t)\, E^\perp\, U(t)\right], \tag{6.2}$$

where $E^\perp = I - E$.

Now consider the following three probabilities for a system prepared in the state described by ρ at $t = 0$:

(1) The probability $P(0, t; \rho)$ that the system is *found* to decay *sometime during the interval* $\Delta = [0, t]$.

(2) The probability $Q(0, t; \rho)$ that no decay is *found throughout the interval* Δ.

(3) The probability $R(0, t_1, t; \rho)$ that the system is *found* undecayed *throughout the interval* Δ but is *found* to decay *sometime* during the subsequent period $[t_1, t] = \Delta_2, 0 < t_1 < t$.

What distinguishes these probabilities from $q(t), p(t)$ is that the latter refer to outcomes of measurements of E at the instant t, *the system being left unobserved* after the initial state preparation until time t, whereas the former probabilities refer to outcomes of *continuously ongoing measurement of E during the entire interval* Δ. The concept of a continuously ongoing observation (or measurement) had not been discussed earlier within the framework of standard orthodox quantum mechanics, and it was not clear in the beginning that it was operationally meaningful. A continuous observation can be regarded as the limiting case of a succession of practically instantaneous measurements as the intervals between successive measurements approach zero. That such a notion can have at least an approximate physical significance is suggested by the cloud and bubble chamber tracks of charged unstable particles. The observation of the tracks amounts practically to continuous monitoring of the existence of the unstable particle during its time of flight through the detection chamber. The first encounter of the incoming charged particle with a vapour molecule in the cloud chamber ionizes it and results in a localized wave packet. If the incident wavelength λ is much smaller than the diameter a of the molecule, the wave packet will propagate in the direction of its initial momentum without appreciable diffraction. This constitutes a position measurement with an uncertainty of the order of a. Now, the width of every localized wave packet becomes appreciably larger than its initial width in a time determined by its initial localization and mass. This is a measure of its Zeno time τ. If the density of the vapour molecules in the cloud chamber is such that the time between successive ionizing encounters is much smaller than τ, the spreading will be inhibited by these repeated interruptions and the track will be linear and in the direction of the initial momentum.

By their definitions the three quantities $P(0, t; \rho)$, $Q(0, t; \rho)$ and $R(0, t_1, t; \rho)$ satisfy the relations

$$P(0, t; \rho) + Q(0, t; \rho) = 1 \qquad (6.3)$$

and

$$R(0, t_1, t; \rho) = Q(0, t_1; \rho) P(0, t - t_1; \rho_1), \qquad (6.4)$$

where ρ_1 is the state in which the system finds itself at time t_1 after being continuously observed and *found* to be undecayed *throughout* $[0, t_1]$. It is therefore sufficient to calculate Q and ρ_1.

We will assume in the spirit of orthodox quantum mechanics that the measurements of E are *ideal* measurements that can be done instantaneously, and that under measurements on the state that yield the result 'yes' or 'undecayed', the state changes according to the projection postulate

$$\rho \to \hat{\rho} = E \rho E. \qquad (6.5)$$

Let us start with the state ρ at time $t = 0$ and make $n + 1$ ideal measurements at times $0, t/n, 2t/n, \ldots, (n-1)t/n$ and t. At each of these measurements the state collapses according to (6.5) but at all intervening times it undergoes Schrödinger time evolution. The (unnormalized) state after collapse is represented by the density matrix $\rho(n, t)$ given by

$$\rho(n, t) = T_n(t)\, \rho\, T_n^*(t) \tag{6.6}$$

where

$$T_n(t) = \left[E\, U(t/n)\, E\right]^n \equiv \left[E \exp(-iHt/n)\, E\right]^n. \tag{6.7}$$

Denoting by $Q(n, \Delta; \rho)$ the probability that the state be found undecayed in *each* of these measurements, one obtains from quantum mechanics the expression

$$Q(n, \Delta; \rho) = \mathrm{Tr}\left[T_n(t)\, \rho\, T_n^*(t)\right]. \tag{6.8}$$

Let us now take the limit $n \to \infty$, and assume that these limits exist. If the limit

$$\lim_{n\to\infty} T_n(t) = \lim_{n\to\infty} \left[E\, U(t/n)\, E\right]^n = T(t) \tag{6.9}$$

also exists for $t \geq 0$, one obtains

$$\rho(t) \equiv \lim_{n\to\infty} \rho(n, t) = \left\{\mathrm{Tr}\left[T(t)\, \rho\, T^*(t)\right]\right\}^{-1} . \, T(t)\, \rho\, T^*(t) \tag{6.10}$$

for the (normalized) state obtained as the result of *continuous observation* and verification that the system remained undecayed *throughout the interval*. The probability of this outcome is given by

$$\begin{aligned} Q(\Delta, \rho) &\equiv \lim_{n\to\infty} Q(\Delta, n; \rho) \\ &= \lim_{n\to\infty} \mathrm{Tr}\left[T_n(t)\, \rho\, T_n^*(t)\right] \\ &= \mathrm{Tr}\left[\rho\, T^*(t)\, T(t)\right]. \end{aligned} \tag{6.11}$$

Therefore,

$$P(\Delta; \rho) = \mathrm{Tr}\left[\rho\, (I - T^*(t))\, T(t)\right]. \tag{6.12}$$

Let us assume continuity of $T(t)$ at $t = 0$ on physical grounds so that

$$\lim_{t\to 0_+} T(t) = E, \tag{6.13}$$

and the probability $Q(\Delta; \rho)$ given by (6.11) approaches the probability $\mathrm{Tr}(\rho E)$ as $t \to 0_+$ that the system is initially undecayed:

$$\mathrm{Tr}(\rho E) = 1. \tag{6.14}$$

This is a desirable requirement for a physical theory and implies that $T^*(t)T(t) = E$ for all t and $T^*(t) = T(-t)$ [151]. It follows then from (6.12) that

$$P(\Delta; \rho) = 0. \tag{6.15}$$

This remarkable result embodies the quantum Zeno paradox – the quantum mechanical probability $P(\Delta; \rho)$ (satisfying (6.12)) that the system prepared in the undecayed state ρ will be observed to decay *sometime* during the interval $\Delta = [0, t]$ vanishes if it is continuously monitored!

The paradoxical result depends critically on the existence of the operator $T(t)$ for $t \geq 0$ and its continuity at $t = 0$. These results can be rigorously proved using a structure theorem concerning semigroups. The reader is referred to the original paper of Misra and Sudarshan [151] for a proof of this theorem.

6.3 Zeno time

We will now give a simple argument based on the uncertainty relations [155] to show how deeply rooted the Zeno paradox is in quantum theory. If A and B are two observables of a system that do not depend explicitly on time, and we take B to be the Hamiltonian, then it follows from the well-known identity

$$\Delta A \, \Delta B \geq \frac{1}{2} |\langle [A, B] \rangle| \tag{6.16}$$

that

$$\Delta A \geq (\hbar/2\,\Delta E)|\,\mathrm{d}\langle A\rangle/\,\mathrm{d}t\,|. \tag{6.17}$$

Choose for A the projection operator $P = |\psi\rangle\langle\psi|$ where $|\psi\rangle$ is the initial undecayed state of the system. (The projection operator E defined above reduces to P if there are no decay products initially.) Then, if $|\phi\rangle = \exp(-iHt)|\psi\rangle$ is the same (undecayed) state at time t due to Schrödinger evolution, we have

$$P(t) = |\langle\phi|\psi\rangle|^2, \tag{6.18}$$

and so

$$(\Delta A)^2 = \langle A^2\rangle - \langle A\rangle^2 = P(1 - P). \tag{6.19}$$

Using this result in (6.17), we finally get

$$[P(1 - P)]^{1/2} \geq (\hbar/2\,\Delta E)|\,\mathrm{d}P/\mathrm{d}t\,|. \tag{6.20}$$

Since $P = 1$ at $t = 0$, it follows that $\mathrm{d}P/\mathrm{d}t$ must vanish at $t = 0$, and hence there cannot be any linear term in t in the expression for $P(t)$. Integration of (6.20) gives

$$P \geq \cos^2\left[(\Delta E)t/\hbar\right], \tag{6.21}$$

which is called Fleming's rule [156]. It shows that the decay probability of the state $(1 - P(t))$ must grow as t^2 and *cannot therefore be exponential*.

This is a rigorous result of quantum theory. The fact that exponential decays *are* observed in nature is due to the availability of an infinite number of vacuum states to the emitted particles in free space. (See the discussion on the modification of spontaneous emission rates, Section 2.2) There have been extensive discussions on deviations from the exponential decay law at very short and long times. (See [155] for a critical discussion of this topic and references to earlier literature.) However, it is very hard to detect such a deviation [157], and none has been observed as yet [158].

It is clear from (6.20) that a *sufficient* condition for observing the Zeno effect is that ΔE (the range of energies accessible to the decay products) is finite. (For a critical discussion on whether it is also *necessary*, see [155].) This implies through the uncertainty relation $\Delta E \, \Delta \tau \geq \hbar/2$ that frequent measurements have to be made within the time interval $\Delta \tau$ in order to observe the Zeno effect. This time interval $\Delta \tau$ is called the Zeno time. The shorter it is, the harder it becomes to observe the effect.

6.4 Experimental observation

That the inhibition of *induced* transitions by frequent interventions offers a much better prospect for experimental observation than the inhibition of *spontaneous* transitions was exploited by Itano *et al.* [154] to carry out an experiment originally proposed by Cook [159] (Fig. 6.1). They used a magnetic trap as a pot to hold several thousand $^9\text{Be}^+$ ions, the 'water'. Trapped and laser-cooled ions provide very clean systems for testing quantum mechanical calculations of transitions. They can be observed for very long periods, free from perturbations and relaxations, and their levels can be manipulated easily with r.f. and optical fields. The three energy levels of ^9Be (Fig. 6.2) that were used for the experiment were: the $(m_I, m_J = (3/2, 1/2)$ and $(1/2, 1/2))$ hyperfine levels in the ground state $2s^2S_{1/2}$ as levels 1 and 2, and the $(m_I = 3/2, m_J = 3/2)$ sublevel of the $2p^2P_{3/2}$ state as level 3. In the beginning all the ions were brought to level 1 by optical pumping for about 5 s. When these ions were exposed to a radio-frequency field in resonance with the transition to level 2 $(E_2 - E_1/\hbar = 320.7$ MHz) for exactly 256 ms (to make it an on-resonance π pulse), all of them moved up to level 2 – the 'water' had boiled *when no one was watching*. Next they decided to repeat the experiment but this time to peep in before 256 ms were up. They used very short pulses (of 2.4 ms duration) of laser light at 313 nm in resonance with the transitions $1 \leftrightarrow 3$ in order to produce strong fluorescence. If the ions are in level 1 at the times of these quick peeps, they would be driven up to level 3, and the consequent fluorescence detected by photon counting. On the other hand, if they are in level 2, they are not available to be excited to level 3 by

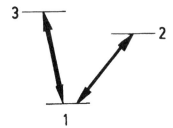

Fig. 6.1.   Energy-level diagram for Cook's proposed demonstration of the quantum Zeno effect (after Ref. [154]).

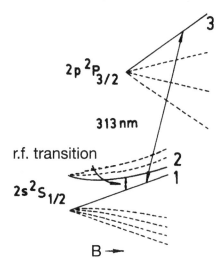

Fig. 6.2.   Diagram of the energy levels of ^{9}Be$^+$ in a magnetic field B. The states labelled 1, 2 and 3 correspond to those in Fig. 6.1 (after Ref. [154]).

the short-duration laser pulses, and no fluorescence would be seen. The fluorescing of level 3 is therefore a very efficient indicator of the survival probability of the ions in level 1. Itano *et al.* found that the number of ions surviving in level 1 when the first laser pulse was sent after 256 ms, was zero. However, as the frequency of these pulses was increased, the survival probability also increased and was almost unity when the pulses were sent every 4 ms, i.e., 64 times in 256 ms. The 'watched pot' had refused to boil, they concluded.

It might be argued that some previous experiments had already indicated the existence of the Zeno effect, such as the spin-flip resonance of a single, trapped electron [160]. The spin state of the electron is detected by coupling it to an electronic circuit. As the interaction time scale is shortened by increasing the coupling, noise broadens the spin-flip resonance, resulting in a decrease of the rate of transitions induced by a weak microwave field. Another example is provided by the three level Hg^+ system that was used to demonstrate interrupted fluorescence (see Chapter 8). The radiation resonant with the $1 \leftrightarrow 3$ transitions was seen to slow down the rate of the $1 \leftrightarrow 2$ transition (induced by a narrow-band laser). In none of these experiments, however, were the intervention times clearly separated from the free, unitary evolution periods. The experiment of Itano *et al.* was the first one to achieve this separation and provide the first clear and unambiguous demonstration of the effect.

As we have seen in Chapter 2 (Section 2.2.2), if an atom is kept in a lossless high-Q resonant microcavity that has no photons in it initially, it will oscillate between the excited and ground states with probabilities $\cos^2 \Omega t$ and $\sin^2 \Omega t$ respectively, where Ω is the coupling constant (the vacuum field Rabi frequency) and t is the time the atom has been in the cavity. Thus the probability for the atom to be in the ground state grows as t^2 for t short compared to Ω^{-1}. This feature will survive if the atom is coupled to a single *damped* cavity mode provided t is short compared to both Ω^{-1} and κ^{-1} ($\kappa = \omega/Q$ being the dissipation rate of the cavity). It might therefore be possible to demonstrate the Zeno effect on the *spontaneous* decay of an atom in a microcavity [154].

A number of papers appeared after the experiment of Itano *et al.* claiming that the experiment did not conclusively demonstrate wave function collapse as claimed, and that the results could be explained on the basis of Schrödinger evolution alone [249], [162], [163], [164], [165]. Wave function collapse is not a universally accepted principle of quantum mechanics but a controversial and debated principle. What makes the Zeno *effect*, which is a general consequence of quantum mechanics independent of its interpretation, a *paradox* is the dramatic effect that the very presence of a macroscopic measuring device seems to have on the time evolution of a system. What the experiment showed, however, can also be interpreted more conventionally as the effect of interactions with an external field. The *paradox* originally envisaged by Misra and Sudarshan therefore remains to be convincingly demonstrated. Inagaki *et al.* [166] have suggested a possible way to demonstrate the paradox that involves repeated neutron spin-flips and measurements. Agarwal and Tewari, on the other hand, have suggested an all-optical experiment with single-photon states of the radiation field to study the quantum Zeno effect. It is based on the optical analogue of the $SU(2)$ dynamical evolution of two-level atoms in external fields [167]. Measurements are simulated by simply opening up various ports and leaving them undisturbed, which induces decoherence – there are no interactions with external fields in their scheme.

6.5   Universality of the Zeno effect

Since the Zeno effect is a general consequence of the principles of quantum mechanics, one would expect it to show up in many areas of physics, whenever the time scales involved are right, irrespective of the details of the dynamics. This is indeed the case. We have already seen some examples such as the existence of cloud chamber tracks and the inhibition of atomic transitions. Another very interesting example is Hund's paradox of optical isomers [168]: sugar molecules occur in two distinct optically active isomeric states, left-handed and right-handed, in spite of the effective Hamiltonian being parity conserving. Let us denote the left-handed and right-handed states of the sugar molecule by $|L\rangle$ and $|R\rangle$ respectively. These states are not eigenstates of the effective parity-conserving Hamiltonian which is expected to induce oscillations between them through its off-diagonal terms. Nevertheless, if one of the two kinds of sugar is prepared, it is found to remain in that state ($|L\rangle$ or $|R\rangle$) for a very long time at low temperatures. Harris and Stodolsky [169] have shown that this stability in a sugar solution can be understood in terms of the 'watched pot effect' resulting from frequent collisions with solvent molecules. They have also pointed out that if the stability persists in vacuum, it could be understood in terms of the 'watchdog effect' arising from the inhibiting influence of weak interactions on the molecules.

The solar neutrino problem in particle physics provides another example of the 'watchdog effect'. Earth based observations indicate that only about $\frac{1}{3}$ of the electron type neutrinos ν_e produced in the sun and expected to reach the earth are detected. Since most people believe that the solar structure is very well known and is unable to account for this depletion, one attractive possibility is the conversion of ν_es to muon type neutrinos ν_μ (which escape detection in the apparatus used) as a result of neutrino oscillations. Such oscillations are theoretically possible if the neutrinos have a small mass. When neutrinos travel through a gas of electrons, there is an additional coherent amplitude for ν_e–electron scattering through the charged weak current (exchange of W^+ bosons), an option not available to the ν_μs because of their different flavour quantum number. This additional interaction can produce a resonant conversion of ν_es to ν_μs in an electron gas of appropriate density. This is known as the MSW effect [170]. Since the electron density inside the sun varies from the core to the surface, this resonance condition can be realized in the sun. This is an example of the 'reverse watchdog effect' in as much as it occurs basically as a result of the continuous coupling of the neutrinos to the solar environment, but here the effect is one of enhancement rather than inhibition and takes place due to resonance between the time scales of vacuum oscillations and the coupling to the environment.

The stability of the neutron inside a nucleus is yet another example of the Zeno effect. The strong-interaction time scale being much shorter than the weak-interaction time scale, the β-decay of the neutron is inhibited inside a nucleus. The influence of one fundamental interaction on another that could come into play is a novel feature of the quantum 'watchdog effect' and could have implications of far-reaching importance.

7

Testing collapse

7.1 Introduction

In spite of its wonderful agreement with every experiment performed so far, standard quantum theory (SQT) fails to describe *events* and to define the circumstances under which such events occur [171]. It is the linearity of Schrödinger evolution that lies, as we have seen, at the root of the problem. A state vector always evolves to become a linear superposition of the states corresponding to several possible outcomes of a measurement, and it is only when an experiment is actually done that one of these possible outcomes is realized at a particular instant. Thereupon the state has to be changed to the one corresponding to the particular outcome in order to follow its subsequent evolution. This additional information is not contained *in the theory* and has to be obtained from *outside*. This means that the theory is unable to predict *when* an event will occur. All it can predict is that *if* an event occurs, the possible outcomes and their probabilities are such and such. Since events do occur in every experiment, there is something missing from the theory.

Since a state vector can be written as the linear sum of a complete set of basis states and these basis states can be chosen in a number of ways, each of which corresponds to a different set of outcomes, the theory also fails to tell us how to choose the *preferred basis*. The various interpretations attempt to supply this missing information. For example, according to the Copenhagen interpretation the measuring apparatus located at a certain position in space determines the preferred basis. However, it does not define what a measuring apparatus is *within the theory*. So one has to go outside the theory to make correct use of it. SQT therefore has two problems, the *events problem* and the *preferred basis* problem.

Furthermore, as we saw in Chapter 5, the linearity of quantum evolution implies macroscopic quantum coherence (MQC) which is yet to be observed. In other words, if one regards SQT as a fundamental theory, the emergence of the classical world is also problematic. Two approaches have been proposed to solve these problems. One is the approach of *decoherence* mentioned in Chapter 5 which retains the linearity of quantum evolution and therefore implies what Pearle calls *False Collapse* [171]. The other is the approach of Ghirardi, Rimini, Weber [172], and Pearle [173] who modify Schrödinger's equation in a well-defined and simple manner so as to retain the approximate agreement with SQT for microscopic systems containing a small number of constituents while describing events paralleling the explicit or *True Collapse* of a state vector. It also automatically explains the emergence of the classical behaviour of macroscopic objects containing a large number of constituents. Bell advertised this approach as being superior to SQT in that it allows a realistic version of quantum theory in which measurements are described by the evolution equations and *no further interpretation is necessary.*

7.2 The GRWP model

There are two versions of this model, one that incorporates spontaneous localization (SL) [172] and the other continuous spontaneous localization (CSL). The evolution of the density matrix $\rho(t)$ in SL is described by the equation

$$\frac{\partial \langle x | \rho | x' \rangle}{\partial t} = -i \langle x | [H, \rho] | x' \rangle - \frac{1}{T} \sum_{k=1}^{N} [1 - \Phi(x_k - x'_k)] \langle x | \rho | x' \rangle$$

(7.1)

where $|x\rangle = |x_1, x_2, \ldots\rangle$ is the position eigenstate for all the particles in the system, $\Phi(z) = \exp(-z^2/4a^2)$ or any other similar function, and T and a are two free parameters. The choice $T \equiv \lambda^{-1} = 10^{16}$ s $\approx 300 \times 10^6$ yr guarantees that the collapse or localization time of a microscopic system is 10^8–10^9 yr. The choice $a = 10^{-5}$ cm. makes the localization distance large compared to typical atomic dimensions and to the mean spreads around the equilibrium positions of the lattice points of a crystal. This avoids the embarrassing occurrence of MQC, i.e., linear superpositions of appreciably different locations of a macroscopic object. On the other hand, the collapse time for a macroscopic object containing $N \simeq 10^{23}$ particles is of the order of 10^{-7} s, a value substantially less than human perception times which are greater than about 10^{-2} s.

The evolution equation in CSL is

$$\frac{\partial \langle x | \rho | x' \rangle}{\partial t} = -i \langle x | [H, \rho] | x' \rangle - \frac{1}{2T} \sum_{j=1}^{N} \sum_{k=1}^{N} [\Phi(x_j - x_k)$$

$$+ \Phi(x'_j - x'_k) - 2\Phi(x_j - x'_k)] \langle x | \rho | x' \rangle. \quad (7.2)$$

It has the advantage of preserving the symmetry of the wave function for identical particles, which the SL model fails to do. Consider a solid macroscopic 'pointer' composed of N particles, each of which is responsive to the collapse mechanism, and suppose that its wave function is initially in a superposition of two states that are separated by a distance greater than a. Then, as we have seen, it will collapse to one of these states in a time

$$\tau_{SL} \simeq T/N \simeq 10^{-7} \text{ s} \qquad (7.3)$$

for $N = 10^{23}$. In the CSL model, however, this collapse time is reduced to

$$\tau_{CSL} = T/NDa^3 \qquad (7.4)$$

where D is the particle number density. Since Da^3 is the average number of particles in the 'collapse volume' which is typically of the order of 10^9 for $N \simeq 10^{23}$, one obtains $\tau_{CSL} \simeq 10^{-16}$ s.

One of the most interesting features of these models and one that makes them empirically falsifiable is the *non-conservation of energy* E due to the appearance of the non-Hamiltonian terms in (7.1) and (7.2) which cause narrowing of wave packets by collapse. For a free particle one obtains for both the SL and CSL models

$$\delta E/t = \frac{3\hbar^2}{4 m a^2 T} . \qquad (7.5)$$

For a microscopic particle of mass $m \simeq 10^{-23}$ g,

$$\delta E/t \simeq 10^{-25} \text{ eV s}^{-1}, \qquad (7.6)$$

which implies that it takes 10^{18} yr for an increase of 1 eV. The same result also holds for the centre of mass of a macroscopic system because the collapse time T decreases and the mass m increases proportional to the number of constituents. However, there is also an increase of energy in the internal motion which is the same for all the constituents. The estimated energy increase for a system of N atoms is

$$\delta E/t \simeq 10^{-14} \quad \text{erg s}^{-1}. \qquad (7.7)$$

The increase in temperature with time for an ideal monatomic gas then turns out to be of the order of 10^{-15} K per year.

The models are therefore able to reproduce SQT results for microscopic objects and classical mechanical results for macroscopic objects, and provide the basis for a conceptually clear description of quantum measurement. They also allow a description of the evolution of macroscopic objects in terms of a classical phase-space density obeying a Markov diffusion process [172]. This class of models gives rise to effects which do not occur in SQT but are in principle observable. Since they have not been seen as yet, there are already limits on the parameters of these models. It is important to know how restrictive these limits are, and how much they can be improved through future experiments. It may turn out that this simple solution to the measurement problem is ruled out on experimental grounds. On the other hand, should these effects turn out to be real, a whole new domain of theoretical and experimental physics would open up.

7.3   Observational tests of GRWP models

7.3.1   *Spontaneous excitation of atoms*

Apart from the increase of the energy of the centre of mass, there is also an increase of energy in the internal motion which is the same for all constituents. Let the normalized ground state wave function of a hydrogen atom be

$$\psi_0 = (\pi a_0^3)^{-1/2} \exp(-r/a_0) \tag{7.8}$$

where a_0 is the radius of the atom. In the GRWP model there is a certain probability that this will spontaneously collapse to the 'hit' state

$$\psi_{0x}^{hit} = (\pi a^2)^{-3/4} R(x)^{-1} \exp\left(-\frac{|x-r|^2}{2a^2}\right) \psi_0, \tag{7.9}$$

where x is the centre about which the collapse occurs and $R(x)$ is chosen so that the hit state is correctly normalized. Since the probability amplitude that the hit state is unexcited by the collapse is given by $\langle \psi_{0x}^{hit} | \psi_0 \rangle$, the probability for a hit to excite the atom is given by

$$\begin{aligned} P &= 1 - \int d^3x \, |R(x)|^2 |\langle \psi_{0x}^{hit} | \psi_0 \rangle|^2 \\ &= 1 - \frac{3a_0^2}{2a^2} + \text{higher orders in } \left(\frac{a_0}{a}\right). \end{aligned} \tag{7.10}$$

It follows therefore that the number of excitations per second per atom is given by

$$\mathcal{N} = \frac{1}{T} \frac{3a_0^2}{2a^2}. \tag{7.11}$$

Using the GRWP parameters $T = 10^{16}$ s, $a = 10^{-5}$ cm and the radius of the hydrogen atom $a_0 = 5 \times 10^{-9}$ cm, one finds the rate [174]

$$\mathcal{N} \simeq 3.75 \times 10^{-23} \, \text{s}^{-1} \tag{7.12}$$

which gives about 20 photons per second from a gramme of hydrogen. To the order in which this result holds only the transitions from the first excited state (the p-state) to the ground state are important, and the photons are largely ultraviolet. The place to look for them would be the large masses of deep underground water used to look for proton decays and neutrinos from supernova explosions. The possible sources of contamination are cosmic rays and excitations caused by thermal motion. The cosmic rays are largely eliminated by going deep underground, and a naive estimate based on the Boltzmann distribution shows that the thermal excitations are many orders of magnitude below the rate suggested above, even at room temperature. Since proton decay seems to be beyond the range of present detection efficiencies and it may be many years before the next supernova explosion occurs, the underground water bodies seem to be the ideal places now to look for the ultraviolet photons coming from collapse.

It has been suggested that the rate of collapse should contain the mass of the particle as a factor [174], which would reduce the photon rate from hydrogen by a factor of 5×10^{-4} or may be even the square of this. Although there are large theoretical uncertainties, it would certainly be worthwhile to determine the upper limit to the number of 'anomalous' photons spontaneously emitted by matter.

7.3.2 Nucleon decay experiments

As is clear from the equations (7.5) and (7.11), an experimental upper limit on the radiation rate from collapse is a restriction on the parameter combination $(Ta^2)^{-1}$:

$$(Ta^2)^{-1}_{\text{expt}} < k(Ta^2)^{-1}_{\text{GRWP}}. \tag{7.13}$$

Since the collapse models deal with a fundamental process, it is reasonable to assume that they apply to all the basic constituents of matter such as electrons and quarks. In that case it is possible to show that the SL model becomes untenable whereas the CSL model can be rescued from marginal tenability *provided the rate of wave function collapse is proportional to the mass of the particles involved* [175], as in the modification of Diosi's model [176] by Ghirardi, Grassi and Rimini [177]. Let $|m_1, m_2, m_3, \ldots\rangle$ be a composite state of a system of particles with m_i the mass in the ith cell of three-dimensional space divided into cells of volume a^3 ($a = 10^{-5}$ cm). Then, according to the CSL model

$$\langle m'_i, m'_j, m'_k, \ldots |\rho(t)| m_i, m_j, m_k, \ldots\rangle = \exp\left[-\frac{1}{2Tm_0^2}\sum_l (m_l - m'_l)^2\, t\right]$$

$$\times \langle m'_i, m'_j, m'_k, \ldots |\rho(0)| m_i, m_j, m_k, \ldots\rangle \tag{7.14}$$

where $\rho(t)$ and $\rho(0)$ are density matrices of the system at time t and $t = 0$ respectively, m_0 is the nucleon mass and $T = 10^{16}$ s is the GRWP parameter. If M is the total mass of the system and D is the mass density (assumed to be uniform), the number of cells with non-zero arbitrary mass $m_i = Da^3$ is M/Da^3. Now consider two non-overlapping states $|m_i, m_j, m_k, \ldots\rangle$ and $|m'_i, m'_j, m'_k, \ldots\rangle$. Then $m_i \neq 0$ implies $m'_i = 0$ and vice versa, and so

$$\sum_i (m_i - m'_i)^2 = 2 \left(\frac{M}{Da^3} \right) (Da^3)^2$$

$$= 2MDa^3 \qquad (7.15)$$

which implies that the off-diagonal elements of the density matrix of the system are exponentially damped within a collapse time scale given by

$$\tau_{CSL} = \frac{Tm_0^2}{MDa^3}. \qquad (7.16)$$

Note that this collapse time does not depend in any way on the separation of the centres of the superposed mass distributions but only on their non-overlapping nature.

With such couplings it is possible to show that the first non-vanishing contribution to the rate of internal excitation of atoms is given [175] by

$$\dot{P}(\phi)_{t=0} \sim \frac{1}{T} \left(\frac{m_e}{m_0} \right)^2 \left(\frac{a_0}{a} \right)^2, \qquad (7.17)$$

where m_e is the electron mass and a_0 the atomic radius. Since m_0 is the nucleon mass, the collapse time for an isolated nucleon ($\sim T(m_0/m)^2$) is unaltered from the GRWP value, as is the centre of mass energy increase of hydrogen gas (7.5). On the other hand, the collapse time for an isolated electron increases by a factor $(m_0/m_e)^2$, and the atomic excitation rate decreases by a factor 10^{-12} compared to the GRWP values. Thus, if the coupling is proportional to the particle mass, the bound state excitation rate is considerably reduced for a given choice of the GRWP parameters a and T.

Let us now turn to the problem of the collapse-induced excitations of quarks in a nucleon. The proton decay experiments have established a lower bound on the proton lifetime of $\sim 10^{31}$ yr $\simeq 3 \times 10^{38}$ s with a 99% confidence level. Since the collapse-induced proton decays would be similar to the decays predicted by Grand Unified Theories (GUT) that the experiments were designed to detect, it is reasonable to assume that the upper limit on the collapse-induced excitation rate has a value of 10^{-38} s^{-1}. For a non-relativistic proton both the SL and the CSL models give, *without any special choice of the coupling constants*, an excitation rate

$$\dot{P}(\phi)|_{t=0} \simeq \frac{1}{T}\left(\frac{a_{\text{nuc}}}{a}\right)^2$$

$$\simeq \frac{(Ta^2)_{\text{GRWP}}}{(Ta^2)} 10^{-32} \text{ s}^{-1}, \tag{7.18}$$

leading to the experimental bound

$$(Ta^2)^{-1} < 10^{-6}(Ta^2)_{\text{GRWP}}^{-1} = 10^{-12} \text{ cm}^{-2} \text{ s}^{-1} \tag{7.19}$$

which is inconsistent with the GRWP choice of parameters. Is it possible to rescue the models by choosing the parameters proportional to the particle masses? After all, the values of the parameters chosen by GRWP [172] were only reasonable guesses, and there is some room for adjustments.

Let us first look at the SL model. If T is increased by a factor 10^6, the collapse time of a 'pointer' containing 10^{23} particles increases to 10^{-1} s, which is larger than the human response time. On the other hand, an attempt to increase a by only a factor of 10 makes it 1 µm, which is larger than a visibly detectable distance. In both these cases collapse would fail to do what it is meant to accomplish. Therefore, the nucleon decay experiments already seem to rule out the SL model of collapse.

In the CSL model, however, the additional factor that reduces τ_{CSL} (7.4) compared to τ_{SL} by a factor of 10^9 allows one to choose T smaller by a similar factor and so bring the model well within the experimental bound (7.19). Nevertheless, the situation is not as comfortable as it might appear. The reason is that proponents of collapse would like to ensure, for example, that even a speck of carbon of radius 10^{-3} cm which is visible to the naked eye but contains only about 10^{15} rather than 10^{23} nucleons, should collapse rapidly enough. But if T is chosen to be even just compatible with the bound (7.19), i.e., if $T = 10^6 T_{\text{GRWP}} = 10^{22}$ s, then according to CSL such a speck should collapse in 10^{-2} s, which is uncomfortably close to human perception time. For other difficulties with the model the reader may like to look at reference [175].

One could, of course, argue that the quarks in a proton are actually relativistic objects, and that relativistic corrections could turn out to be significant. Unfortunately, it is very difficult to construct relativistic collapse models that are satisfactory [178]. The interesting point is that if one makes some reasonable assumptions, future experiments on anomalous photon emission from matter as well as the proton decay experiments already carried out are able to put bounds on the parameters that occur in collapse models of the GRWP genre, rule out some of these models and can even indicate whether these parameters are proportional to mass or energy, suggestive of a gravitational effect.

7.3.3 Cosmological tests

It turns out that apart from laboratory experiments, cosmological observations can also constrain the parameters of collapse models. Although general relativity is crucial for a proper understanding of astrophysics and the physics of the early universe, and no satisfactory relativistic generalizations of the collapse models are yet available, nevertheless it is instructive to use one of the key results of these collapse models that is unlikely to go away in a relativistic version, namely a small rate of energy non-conservation, to see if it leads to a significant effect when integrated over a cosmological time scale. Such a calculation has been done by Home and Majumdar [179] using the CSL model which, as we have seen, can be barely rescued by making the parameters mass dependent.

Let us calculate the total amount of energy liberated in the universe up to the present era due to continuous spontaneous localization of the baryonic wave function. In order to do this, let us recall that the time evolution of the scale factor $R(\tau)$ in a flat Friedman–Robertson–Walker (FRW) cosmological model is given by

$$\frac{\dot{R}^2}{R^2} = 8\pi\, G\, \rho_{\text{tot}} \tag{7.20}$$

where the derivatives are with respect to the FRW time frame τ and ρ_{tot} is the total energy density of the universe consisting of contributions due to radiation ρ_{r}, matter ρ_{m}, the cosmological constant ρ_λ *and* the CSL contribution ρ_{CSL}. Now, according to the currently popular scheme of electroweak baryogenesis, the baryons are supposed to have been formed during the period of the electroweak symmetry-breaking phase transition in the early universe. It is then reasonable to assume that the CSL process of baryonic wave function collapse and the consequent creation of energy started during this period and continues right up to the present era. Since no baryons could have been created after the period of baryogenesis was over, it is also reasonable to assume that a major part of the energy created by the CSL process produced non-baryonic 'dark' matter. Now, the rate of change of the energy density of the universe due to matter is given by

$$\frac{\mathrm{d}}{\mathrm{d}\tau}\left(\rho_{\text{m}}\, R^3\right) = \frac{\mathrm{d}}{\mathrm{d}\tau}\left(\rho_{\text{CSL}}\right) R^3 . \tag{7.21}$$

In standard cosmology without CSL the right-hand side vanishes, but in the presence of CSL it does not. If the CSL parameters are taken to be mass dependent, and one neglects the lepton masses compared to the baryon masses, one obtains

$$\frac{\mathrm{d}\,\rho_{\mathrm{CSL}}(\tau)}{\mathrm{d}\tau} = \frac{3\,\hbar^2\,m(\tau)}{4\,m_0^2\,a^2\,T} \tag{7.22}$$

for the rate of energy increase due to baryons alone, the additional factor $m(\tau)/m_0$ compared to (7.5) appearing because of the mass dependence, $m(\tau)$ being the mass density of the baryons and m_0 the proton mass. If one takes the proton mass to be the reference mass for all baryons, one has $m(\tau) = m_p n(\tau)$ where $n(\tau)$ is the baryon number density at time τ. It then follows from (7.21) and (7.22) that

$$\dot\rho_{\mathrm{m}} = \frac{3\,\hbar^2\,n(\tau)}{4\,m_{\mathrm{p}}\,a^2\,T} - 3\,m_{\mathrm{p}}\,n(\tau)\,\frac{\dot R}{R}. \tag{7.23}$$

Integrating this equation up to the present time τ_0 using the standard results of cosmology regarding the behaviour of the number density $n(\tau)$ and the scale factor R at different eras, one obtains the total contribution of matter to the present energy density of the universe to be [179]

$$\rho_{\mathrm{m}}(\tau_0) \simeq \frac{2\,n(\tau_0)\,\tau_0^2\,Z}{(\tau_{\mathrm{EQ}}\,\tau_{\mathrm{EW}})^{1/2}} \tag{7.24}$$

where $Z = 3\hbar^2/4\,m_{\mathrm{p}}\,a^2\,T$, τ_{EQ} is the time up to which matter-radiation equality holds and τ_{EW} is the time of electroweak phase transition. Substituting the numerical values of all the parameters [180], one obtains the remarkable result

$$\rho_{\mathrm{m}}(\tau) \simeq 10^{-29}\ \mathrm{g\ cm}^{-3} \simeq \rho_{\mathrm{c}} \tag{7.25}$$

where ρ_{c} is the critical density required for a flat FRW universe. The observed matter density in the universe is at best only 20% of this critical value. Astronomical observations such as rotation velocities of galaxies, on the other hand, indicate that there is at least an order of magnitude more non-luminous or 'dark' matter than luminous matter in the universe, which should make the universe flat. Inflationary cosmology also predicts a flat universe and consequently the existence of a large quantity of dark matter in the universe [181]. One of the outstanding problems in cosmology is to account for the source of this dark matter. It is remarkable that a fundamental process like CSL should be able to account for it. Of course, this could simply be fortuitous. The result (7.24) depends critically on τ_{EQ} which can increase or decrease depending on the amounts of energy and dark matter created by CSL, spoiling the agreement. So it is not possible to draw any reliable conclusion without doing a more detailed calculation. Nevertheless, the interesting lesson is that cosmological bounds on the energy density in the universe can be used to constrain the parameters of CSL and similar models, and that the rate of energy creation by CSL seems to be consistent with such bounds.

7.4 Distinguishing true collapse from environmentally induced decoherence (EID)

As we have seen in Chapter 5, there are certain approaches in which true collapse does not actually occur but the environment of the system to which it is coupled induces the reduced density matrix of the system to become effectively diagonal after a so-called decoherence time [182], [117], [118]. It has been argued that such EID schemes do not satisfactorily solve the problems of SQT [171]. They can, in fact, be empirically distinguished from true collapse models such as CSL, at least in principle. Let us consider a simplified version of the full CSL model due to Pearle [171]. Let the normal Schrödinger evolution produce the pure state

$$|\psi, 0\rangle = \alpha |a\rangle + \beta |b\rangle \tag{7.26}$$

where $|a\rangle$ and $|b\rangle$ are macroscopically different states that are eigenstates of some operator A with eigenvalues a and b respectively (such as different 'pointer positions'), and $|\alpha|^2 + |\beta|^2 = 1$. Let the subsequent evolution proceed according to the modified equation

$$\frac{d}{dt} |\psi, t\rangle_w = -i H |\psi, t\rangle_w - \frac{1}{4\lambda} [w(t) - 2\lambda A]^2 |\psi, t\rangle_w \tag{7.27}$$

where $w(t)$ is a white-noise function and λ is the collapse rate. Every choice of $w(t)$ gives a different evolution, which is indicated by the subscript w attached to the state vector. Since (7.27) is non-unitary, the norm of the state vector changes with time. In order that the largest norm states are the most probable, the following probability rule is also stipulated:

$$\text{Prob}\{w(t) \text{ for } 0 \leq t \leq T\} = Dw \, _w\langle \psi, t | \psi, t \rangle_w \tag{7.28}$$

where Dw is the functional integral element

$$Dw \equiv \Pi_{t=0}^{t=T} \frac{dw(t)}{\sqrt{2\pi \lambda (dt)^{-1}}}. \tag{7.29}$$

One may also assume that $H = 0$ as it only affects the internal evolution of the pointer states. Then

$$|\psi, T\rangle_w = F\{w(t)\} \left\{ \alpha |a\rangle e^{-K(\lambda, T, a)} + \beta |b\rangle e^{-K(\lambda, T, b)} \right\} \tag{7.30}$$

is a solution to (7.27) with the initial condition (7.26) and

$$K(\lambda, T, a) = \frac{1}{4\lambda T} [B(T) - 2\lambda T a]^2$$

$$K(\lambda, T, b) = \frac{1}{4\lambda T} [B(T) - 2\lambda T b]^2$$

$$B(T) \equiv \int_0^T \mathrm{d}t \, w(t). \tag{7.31}$$

The probability that $B(T)$ lies between B and $B + \mathrm{d}B$ is obtained by taking the scalar product of (7.30) with itself and integrating over $w(t)$ for all t except T:

$$\mathrm{Prob} = \frac{\mathrm{d}B}{\sqrt{2\pi\lambda T}} \left\{ |\alpha|^2 \, e^{-2K(\lambda, T, a)} + |\beta|^2 \, e^{-2K(\lambda, T, b)} \right\} \tag{7.32}$$

This shows that as T increases, the state vector (7.30) asymptotically reduces to either $|a\rangle$ with the standard SQT probability $|\alpha|^2$ or to $|b\rangle$ with the standard SQT probability $|\beta|^2$, the only probable values of B lying within a few standard deviations of $2\lambda T a$ in the former and $2\lambda T b$ in the latter case. Thus, a simple and well-defined modification of Schrödinger's evolution is able to describe events paralleling the collapse of the state vector while retaining the good agreement with SQT.

Let us finally write down the density matrix corresponding to the state (7.30):

$$\rho(T) = \int Dw_w \langle \psi, T | \psi, T \rangle_w \frac{|\psi, T\rangle_w \, {}_w\langle \psi, T|}{{}_w\langle \psi, T | \psi, T \rangle_w}$$

$$= |\alpha|^2 |a\rangle\langle a| + |\beta|^2 |b\rangle\langle b|$$

$$+ [\alpha\beta^* |a\rangle\langle b| + \alpha^*\beta |b\rangle\langle a|] \, e^{-\frac{\lambda T}{2}(a-b)^2}. \tag{7.33}$$

The off-diagonal elements of this density matrix decay exponentially because the state vectors corresponding to different $B(T)$s collapse at different rates.

Consider now what happens in a typical environment-induced decoherence scheme. The state vector (7.26) evolves according to the unmodified Schrödinger equation

$$\frac{\mathrm{d}}{\mathrm{d}t} |\psi, t\rangle_w = -\mathrm{i}\, w(t) A |\psi, t\rangle_w \tag{7.34}$$

where $w(t)$ is a sample function of white noise representing the influence of the bath to which the system is coupled. This means that repeated evolutions differ in $w(t)$ whose probability of appearance is guided by the rule

$$\mathrm{Prob}\{w(t) \text{ for } 0 \le t \le T\} = Dw \exp\left[-\frac{1}{2\lambda} \int_0^T \mathrm{d}t \, w^2(t)\right]. \tag{7.35}$$

The state (7.26) therefore evolves to

$$|\psi, T\rangle_w = \alpha\, e^{-\mathrm{i}B(T)a} |a\rangle + \beta\, e^{-\mathrm{i}B(T)b} |b\rangle. \tag{7.36}$$

This shows that *the state vector remains a pure state and does not collapse.* So, no event can be said to have occurred. Nevertheless, it turns out that the density matrix corresponding to this state is still given exactly by (7.33)! This feature is typical of decoherence schemes and occurs in more realistic decoherence schemes as well [171]. The proponents of decoherence schemes therefore claim that after a sufficiently long time when the off-diagonal elements of the density matrix can be effectively ignored (the so-called decoherence time) , an event (*a* or *b*) occurs for the system. This claim is seen to be clearly untenable if one looks at the pure state vector (7.36) which gives rise to the density matrix (7.33). In other words, although the density matrices look similar, they actually describe different physics.

In fact, Pearle has argued that one should not use the phrase 'decoherence time' at all since it seems to imply that there is a physical process called 'decoherence' which takes a certain time to occur. He prefers to use the acronym NOWEN time, NOWEN standing for 'No One Will Ever kNow'. For example, suppose someone (wrongly!) claims that an event takes place at time T_e, and someone else performs an experiment to test this claim, thereby actually causing events to occur. The NOWEN time is, by definition, the time at which the off-diagonal elements of the density matrix become so small that they are judged to be beyond experimental detection. Only if one chooses T_e to be equal to or greater than the NOWEN time would the experimental result be the same *as if* the claim was right. With this choice, no one will ever know that the claim was wrong! On the other hand, if one chooses T_e to be smaller than the NOWEN time, the discrepancy between the different predictions of the two density matrices characterized by different collapse times (both denoted by the same symbol λ above) would enable one to test whether true collapse actually occurs.

7.4.1 *A* gedanken *example*

Consider, for example, the *gedanken* experiment proposed by Home and Bose [183] who consider a version of CSL in which the rate of collapse is proportional to the particle mass involved. Let a pulse of thermal neutrons (wavelength $\lambda \sim 1\,\text{Å}$) be incident on the beam splitter B (Fig. 7.1). One of the sub-beams enters the system M of four rigidly connected mirrors M_1, M_2, M_3 and M_4 which can slide up and down together in the Y-direction. The beam first traverses the M_1–M_2 subsystem, displacing it (and thus M) in the Y-direction by an amount d_R, gets reflected by a fixed mirror F_1, then traverses the M_3–M_4 subsystem, this time displacing M downward by the distance d_R, and finally emerges from M. The other sub-beam is simply reflected by the fixed mirror F_2. The two sub-beams

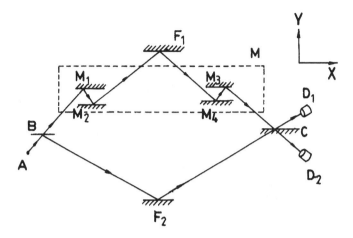

Fig. 7.1.   Sketch of the *gedanken* experiment (after Ref. [183]).

then recombine on the screen D. Let the state of the neutron pulse after passing the beam splitter B be given by

$$|\psi\rangle = C_1|\psi_1\rangle + C_2|\psi_2\rangle \tag{7.37}$$

where $|\psi_1\rangle$ and $|\psi_2\rangle$ are the states of the two sub-beams. After one of the sub-beams has traversed the M_1–M_2 system, the state of the combined system of the neutron and the mirror system M at, say, time t_1 is given by

$$|\phi(t_1)\rangle = C_1|\psi_1\rangle|y_0 + d_R\rangle + C_2|\psi_2\rangle|y_0\rangle \tag{7.38}$$

where $|y_0\rangle$ is the initial state of M and $|y_0 + d_r\rangle$ is its state at time t_1. Let us assume that the displacement of M is macroscopic by some agreed criterion. Then $\langle y_0 + d_R | y_0\rangle = 0$. If we assume that the mass associated with the neutron pulse is negligible in comparison with the mass of the mirror system M, the neutron pulse does not lose any significant energy on traversing M_1–M_2, and so at a later instant t_2 it is able to displace M_3–M_4 (and thus M) downward by exactly the same amount as the earlier upward displacement d_R. The state of the combined system is therefore given at time t_2 by

$$|\phi(t_2)\rangle = (C_1|\psi_1\rangle + C_2|\psi_2\rangle)\,|y_0\rangle. \tag{7.39}$$

This shows that the coherence between $|\psi_1\rangle$ and $|\psi_2\rangle$ is restored. SQT therefore predicts that interference effects between them should be observable on D.

On the other hand, if the displacement d_R is macroscopic, then the state (7.38) should collapse to a mixture or become effectively decoherent according to both CSL and EID models provided $(t_2 - t_1) > \tau_{CSL}$ or τ_{EID} where τ_{CSL} and τ_{EID} are the characteristic time scales in these models.

These models therefore predict that no interference effects between $|\psi_1\rangle$ and $|\psi_2\rangle$ should be observable on D provided $(t_2 - t_1) > \tau_{CSL}$ or τ_{EID}. This can therefore be used to distinguish SQT from CSL and EID models. One can, in fact, go further and distinguish between CSL and EID models by exploiting the difference between τ_{CSL} and τ_{EID} and choosing $(t_2 - t_1)$ appropriately.

As we have seen, the collapse time τ_{CSL} for a non-overlapping mass distribution is given by (7.16) which does not depend at all on the magnitude of the spatial separation between the superposed states. In EID models, on the other hand, the system is coupled to its environment through an interaction Hamiltonian proportional to its position observable. The density matrix of the combined system traced over the unobservable degrees of freedom of the environment evolves according to the equation

$$\langle x | \rho(t) | x' \rangle = \exp\left[-\frac{2M\lambda K \theta (x - x')^2}{\hbar^2} t \right] \langle x | \rho(0) | x' \rangle \qquad (7.40)$$

where x and x' are the points where the superposed wave packets peak, θ is the temperature of the environment and λ^{-1} is the relaxation time for the system whose maximum value is $\simeq 10^{17}$ s. The decoherence time τ_{EID} calculated from (7.40) is

$$\tau_{EID} = \frac{\hbar^2}{2M\lambda K \theta (x - x')^2} \qquad (7.41)$$

which is extremely sensitive to the spatial separation $(x - x')$ of the centres of the superposed wave packets.

Now, if the mirror doublets M_1–M_2 and M_3–M_4 are separated by about 10 cm, thermal neutrons would typically take about 10^{-4} s to travel between them. Thus, provided one can choose both τ_{CSL} and τ_{EID} $\simeq 10^{-5}$ s, it should be possible to distinguish SQT from CSL and EID models. This can be done by suitably choosing the mass of the mirror system M to be $10^{12} m_0$ and arranging the neutron pulses to consist of about 10^8 neutrons to produce a displacement d_R of M of about 10^{-3} cm (with $\theta \simeq 10^2$ K) [183].

To discriminate between CSL and EID models, one has to increase τ_{EID} to about 10^{-3} s, which can be done according to (7.41) by choosing $d_R = 10^{-4}$ cm which is larger than the collapse length a. However, τ_{CSL} still remains $\simeq 10^{-5}$ s because it does not depend on the separation d_R at all. The time of flight of the neutron pulses also remains more or less 10^{-4} s, which is intermediate between τ_{CSL} and τ_{EID}. Under these conditions EID does not produce any appreciable decoherence whereas CSL predicts collapse.

The practical problem of doing an experiment of this kind is to design the mirror system of an appropriate size and mass and to focus the neutron pulses on a very small area of the mirrors. Even if an actual experiment can be done to rule out certain versions of CSL or EID, it would always be possible to formulate variants that would escape falsification. Nevertheless, the usefulness of such an experiment would lie in its ability to restrict the options available.

7.4.2   *An example from biochemistry*

A curious example of the measurement paradox has been recently discussed [184] which highlights the difficulties inherent in the various approaches to the measurement problem. Since DNA molecules occupy a strategic position between microscopic and macroscopic bodies, and the quantum mechanics of macromolecules has provided new results of great significance to molecular biology (such as the stability of the DNA, the folding of proteins to form characteristic globular shapes, the stereoscopic recognition of a substrate by an enzyme, etc.), they should be describable, at least in principle, by a quantum mechanical wave function. The idea is to see if the emission of a single γ photon within a certain time interval, a quantum event, can be detected by a mesoscopic system such as a DNA molecule. The single photon is first passed through a caesium iodide crystal to produce a transient shower of about 10^6 uv photons of wavelength around 250–300 nanometers. This photon shower is then passed through an aqueous solution containing a sufficient number of DNA molecules and an enzyme known as photolyase. This ensures that at least one uv photon is absorbed by a DNA molecule, which undergoes a microscopic but stable change of shape in such a way that molecules of photolyase invariably bind to it. This makes it possible to inspect and identify the damaged DNA molecule at any later time at the discretion of the experimenter. This is therefore an example of a quantum event producing a macroscopically measurable and therefore classical effect through the chemical changes it induces. However, according to unitary Schrödinger evolution the DNA molecule should continue to remain in a superposition of its damaged and undamaged states until it is *observed*. In other words, the process should produce mesoscopic Schrödinger cat states. But, since an individual DNA molecule acts like a switch – it is either 'on' and attracts the enzyme *or* 'off' and does not attract the enzyme – it cannot be in a Schrödinger cat state. To what extent do the unorthodox approaches to the measurement problem ensure this?

Since the uv-induced changes of shape are of the order of 2–3 Å, they are much smaller than the localization scale a ($\approx 10^{-5}$ cm) used in explicit collapse models. Any attempt to reduce this scale significantly would lead to an unacceptably large spontaneous creation of energy. It is not clear therefore how explicit collapse models can take account of this.

The EID models do no better. The estimated decoherence times [182] are calculated on the assumption that the superposed wave packets in position space are separated by distances much larger than the thermal de Broglie wavelength ($\lambda_d = h/\sqrt{2m\pi k_B T}$ where m is the mass of the system, T is the temperature of the environment and k_B is the Boltzmann constant). If the separations are of the same order as λ_d or smaller, there is no decoherence [185]. Now, for a typical synthetic circularized DNA fragment consisting of 500 base pairs, $\lambda_d \simeq 10^{-10}$ cm which is much smaller than the uv-induced displacements ($\simeq 10^{-8}$ cm). However, it is possible to make smaller DNA molecules for which $\lambda_d \simeq 10^{-8}$ cm, but then the estimation of decoherence time becomes ambiguous.

In the de Broglie–Bohm interpretation, as we have seen, the *position* of a 'particle' plays as fundamental a role as the wave function, and provides an ontological description of quantum measurements. One might therefore argue that although the wave packets describing the damaged and undamaged DNA molecule have an appreciable spatial overlap, the actual *positions* of the damaged sites do not overlap and can be recognized by the photolyase enzymes which attach themselves to such sites. However, it is necessary for the wave packets not to overlap if a proper measurement is to take place even in the de Broglie–Bohm interpretation. Therefore, an additional assumption ascribing a more fundamental role to *position* than is implied by the mathematical formulation of the Bohmian scheme seems to be necessary if it is to explain this example.

Thus, if one believes that in a satisfactory physical theory an event should occur dynamically and independent of any conscious intervention by an observer, biomolecular versions of Schrödinger cats such as the one just discussed show that all the non-orthodox versions of quantum mechanics proposed so far require fairly serious refinements.

8

Macroscopic quantum jumps

8.1 Introduction

The quantum theory of atoms and molecules had its origin in the famous 1913 paper of Bohr [186] in which he suggested that the interaction of radiation and atoms occurred with the transition of an atom from one stationary internal state to another, accompanied by the emission or absorption of radiation of a frequency determined by the energy difference between these states. These transitions came to be known as 'quantum jumps'. Since all experiments until the late 1970s used to be carried out with large ensembles of atoms and molecules, these jumps were masked and could not be directly observed; they were only inferred from spectroscopic data. In fact, with the advent of quantum mechanics they eventually came to be regarded as artefacts of Bohr's simple-minded semi-classical model. But with the availability of coherent light sources and single ions prepared in ion traps [187], [188], and optically cooled [189], [190], the issue has been reopened with the experimental demonstration of quantum jumps in single ions [191], [192], [193]. These experiments have also opened up the contentious issue of wave function collapse or reduction on measurement, as they can be regarded as making collapse visible on the oscilloscope screen [191].

Evidence of the discrete nature of quantum transitions in a single quantum system had been accumulating from the observation of photon anti-bunching in single-atom resonance fluorescence [194], the tunneling of single electrons in metal–oxide–semiconductor junctions [195] and spin-flips of individual electrons in a Penning trap [196]. However, what distinguishes the experiments since 1986 [191], [192], [193] from the earlier ones is the direct observation of discrete jumps in optical transitions in a single ion. We will now sketch a simple theory of these effects [197], [198].

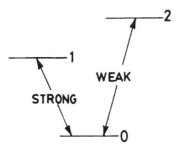

Fig. 8.1. Energy-level scheme for the single-atom double-resonance experiment proposed by Dehmelt (after Ref. [197]).

8.2   Theory

The idea was originally suggested by Dehmelt [199] in the context of detecting a weak transition in single-atom spectroscopy. Fluorescence from a strong optical transition ($\approx 10^8$ photons/s) in a single atom caught in a radio-frequency trap can be easily detected either visually (using a microscope) or photoelectrically, but a weak transition (≈ 1 photon/s) is hard to detect directly. Dehmelt suggested the double-resonance scheme illustrated in Fig. 8.1. Suppose a weak transition $0 \leftrightarrow 2$ and a strong transition $0 \leftrightarrow 1$ share a common lower level 0. If the atom is prepared in the state 0 and irradiated by a laser tuned to the $0 \leftrightarrow 1$ transition, the electron will undergo rapid transitions between the two states, and the rate of fluorescence (photons/s) on this transition will be $R_1 = A_1 P_1 = A_1/2$ since the probability P_1 of the electron to be in the state 1 is $\frac{1}{2}$ (A_1 being the Einstein spontaneous emission coefficient for the strong transition). If a detector is placed at a distance r from the atom, the intensity I_0 of scattered radiation detected by it will be proportional to R_1 and inversely proportional to r^2. This radiation will, of course, not be continuous but will consist of a series of pulses as the photons arrive. However, as the emission rate of photons is very high for the strong transition, this discreteness may be ignored.

Now let the atom be simultaneously subjected to a second radiation tuned to drive the weak transition $0 \leftrightarrow 2$. The moment the electron makes a transition to level 2, the fluorescence intensity I_0 will drop to zero because the electron is no longer available for the strong transition – it is 'shelved' in level 2 which is assumed to be metastable. The fluorescence will turn on again when the electron makes a spontaneous or induced transition back to level 0. Since the weak transitions will occur randomly in time, the atomic fluorescence intensity $I(t)$ will be randomly interrupted and will have the form of a random telegraphic signal (Fig. 8.2). Thus,

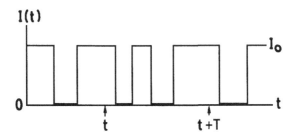

Fig. 8.2. Single-atom fluorescent intensity versus time. Interruptions of fluorescence are due to excitation of the weak transition $0 \rightarrow 2$ (after Ref. [197]).

the interruptions of the fluorescence intensity generated by the strong transition are direct indicators of the weak transition – the fluorescence is on when level 2 is empty and off when it is occupied.

We will first deal with the case of incoherent excitation [197]. The Einstein rate equations are

$$\dot{P}_1 = -A_1 P_1 + B_1 U_1 (P_0 - P_1), \tag{8.1}$$
$$\dot{P}_2 = -A_2 P_2 + B_2 U_2 (P_0 - P_2), \tag{8.2}$$

where P_0, P_1 and P_2 are the probabilities that the levels $0, 1$ and 2 are occupied, and A_i and B_i $(i = 1, 2)$ are the Einstein coefficients for the strong and weak transitions (with $A_1 \gg A_2$). Since

$$P_0 + P_1 + P_2 = 0, \tag{8.3}$$

no separate rate equation is required for P_0. The *mean* rates of fluorescence (photons/s) on the two transitions are $R_i = A_i P_i (i = 1, 2)$.

In the situation in which the strong transition is saturated ($U \rightarrow \infty$), the levels 0 and 1 will be equally populated ($P_0 = P_1$). It is then convenient to introduce the new variables

$$\mathscr{P}_+ = P_2,$$
$$\mathscr{P}_- = P_0 + P_1, \tag{8.4}$$

which are respectively the probability that the weak transition is excited and not excited. In terms of these variables the Einstein rate equations (8.1), (8.2) and (8.3) take the form

$$\dot{\mathscr{P}}_+ = -R_- \mathscr{P}_+ + R_+ \mathscr{P}_-, \tag{8.5}$$
$$\dot{\mathscr{P}}_- = R_- \mathscr{P}_+ - R_+ \mathscr{P}_-, \tag{8.6}$$
$$\mathscr{P}_+ + \mathscr{P}_- = 1, \tag{8.7}$$

where $R_+ = \frac{1}{2} B_2 U_2$ and $R_- = A_2 + B_2 U_2$. These equations describe effectively a two-level system with upward transition rate R_+ and downward rate R_-. In the steady state $\dot{\mathscr{P}}_+ = \dot{\mathscr{P}}_- = 0$, and the solution to (8.5), (8.6) and (8.7) is

$$\mathscr{P}_+ = \frac{R_+}{(R_+ + R_-)}, \tag{8.8}$$

$$\mathscr{P}_- = \frac{R_-}{(R_+ + R_-)}. \tag{8.9}$$

This solution does not describe the stochastic processes actually involved in the transitions, but determines only the mean value of the telegraphic signal ($\bar{I} = I_0 \mathscr{P}_-$) in the steady state. To obtain the fluctuations which are of real interest in this case, one has to take into account the stochastic processes in greater detail. Let $P_{n,\pm}(t, T)$ denote the probabilities that n transitions occur between levels $+$ and $-$ during the time interval $[t, t+T]$ and that the atom is left in level $\pm$ at time $t+T$. Then the probability that altogether n transitions occur between levels $+$ and $-$ during the interval $[t, t+T]$ regardless of the final state is $P_n(t, T) = P_{n,+}(t, T) + P_{n,-}(t, T)$. Thus the probabilities for the final states $+$ and $-$, regardless of the number of transitions, are

$$\mathscr{P}_+(t + T) = \sum_{n=0}^{\infty} P_{n,+}(t, T), \tag{8.10}$$

$$\mathscr{P}_-(t + T) = \sum_{n=0}^{\infty} P_{n,-}(t, T). \tag{8.11}$$

If the stochastic process is stationary, the probabilities $P_{n,\pm}(T)$ are functions of T only, and $\mathscr{P}_\pm$ reduce to constants.

In order to calculate the probability densities for the duration T of 'on times' and 'off times' fluorescent signals, one has to write down the rate equations for the probabilities $P_{n,\pm}$ and solve them. A simple way to do that is to make use of a diagram. In Fig. 8.3 the filled circles represent the probabilities $P_{n,\pm}(T)$ that n transitions occur, one at a time ($n \rightarrow n+1$), in time T and that the atom is left in the state $\pm$. The arrows indicate the flow of probability as T increases, the up arrows representing the rate R_+ and the down arrows the rate R_-. An inspection of the diagram shows that the rate equations are

$$\frac{dP_{n,+}}{dT} = R_+ P_{n-1,-} - R_- P_{n,+}, \tag{8.12}$$

$$\frac{dP_{n,-}}{dT} = R_- P_{n-1,+} - R_+ P_{n,-}. \tag{8.13}$$

It is obvious from the diagram that no probabilities feed $P_{0,\pm}$. Therefore, the first term on the right-hand side of these equations for $n = 0$ must be absent, i.e., $P_{-1,\pm} = 0$. To check that these equations are correct, sum them

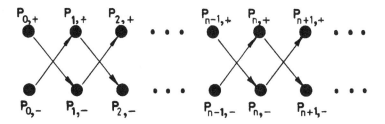

Fig. 8.3.   Filled circles represent probabilities $P_{n,\pm}(T)$ that n transitions occur in time T and that the atom is left in state $\pm$. Arrows indicate the flow of probability as T increases (after Ref. [197]).

over all values of n and use (8.10) and (8.11): the Einstein rate equations (8.5) and (8.6) are recovered.

In order to solve these equations, one needs to specify the initial conditions. Since no transition can occur in a zero time interval, one must have $P_{n,\pm}(t, 0) = 0$ for $n > 0$. It then follows from (8.10) and (8.11) that $P_{0,\pm} = \mathscr{P}_{\pm}(t)$. This completes the specification.

Suppose now that the fluorescence turns off at some time t, signalling that the electron has just made a transition to the upper level (2). Then, $\mathscr{P}_{+}(t) = 1$ and $\mathscr{P}_{-}(t) = 0$, and we have the initial condition $P_{0,+}(t, 0) = \mathscr{P}_{+}(t) = 1$. The solution to (8.12) for $n = 0$ is

$$P_{0,+}(t, T) = \exp(-R_- T) \tag{8.14}$$

which is the probability that no transitions have occurred in time T and the fluorescence is still off. Therefore, the probability that the fluorescence is turned on in time T is given by $F(T) = 1 - \exp(-R_- T)$. The derivative of this distribution, $W_{\text{off}} = \mathrm{d}F/\mathrm{d}T$, is the probability density for the time interval T of 'off times' or interruptions in the fluorescent signal:

$$W_{\text{off}} = R_- \exp(-R_- T). \tag{8.15}$$

Similarly, the probability density for the time interval T of 'on times' is

$$W_{\text{on}} = R_+ \exp(-R_+ T). \tag{8.16}$$

Since $A_2 = R_- - 2R_+$, the Einstein A coefficient for the weak transition can be measured from the values of R_+ and R_- obtained from a statistical analysis of the random signal $I(t)$.

Let us now consider the two-time correlation function

$$C(T) = \langle I(t)I(t + T) \rangle, \tag{8.17}$$

which is a function of T only if $I(t)$ is stationary. It is clear from Fig. 8.2 that the product $I(t)I(t + T)$ can only have the values I_0^2 (when the fluorescence is on at time t as well as at time $t + T$) and zero (at all other times). Therefore,

$$C(T) = I_0^2 P_{on, on}(T),$$ (8.18)

where

$$P_{on, on}(T) = \sum_{n \text{ even}} P_{n, -}(T)$$ (8.19)

is the probability that the fluorescence is on at both times t and $t + T$, $P_{n, -}(T)$ being the probability that the atom is in the level $-$ after n transitions between t and $t + T$. The sum over only even values of n occurs because the signal will be on at both t and $t + T$ only provided an even number of transitions occurs during this time interval. A similar argument shows that the total probability that the signal is on at time t and off at time $t + T$ is

$$P_{on, off}(T) = \sum_{n \text{ odd}} P_{n, +}(T).$$ (8.20)

We will now derive the coupled rate equations for these probabilities. Summing (8.12) over even values of n, (8.13) over odd values of n and making use of (8.19) and (8.20), one obtains

$$\frac{dP_{on, on}}{dT} = R_- P_{on, off} - R_+ P_{on, on},$$ (8.21)

$$\frac{dP_{on, off}}{dT} = R_+ P_{on, on} - R_- P_{on, off}.$$ (8.22)

It follows from (8.19) and (8.20) that in steady state the initial conditions are $P_{on, on}(0) = \mathscr{P}_- = R_-/(R_+ + R_-)$ and $P_{on, off}(0) = 0$. With these conditions the solution to (8.21) and (8.22) for $P_{on, on}(T)$ is

$$P_{on, on}(T) = \frac{R_-^2}{R^2} + \frac{R_+ R_-}{R^2} e^{-(R_+ R_-)T},$$ (8.23)

where $R = R_+ + R_-$. Using this result in (8.18), one obtains

$$C(T) = m_I^2 + \sigma_I^2 e^{-RT},$$ (8.24)

where $m_I = \langle I \rangle = I_0 R_-/R$ is the mean intensity and $\sigma_I^2 = \langle I^2 \rangle - \langle I \rangle^2 = I_0^2 R_+ R_-/R^2$ is the variance of intensity. This shows that the two-time correlation function falls exponentially from the mean square value $\langle I^2 \rangle$ to the squared mean value $\langle I \rangle^2$ as T goes from zero to infinity. Another quantity of interest is the frequency spectrum

$$S(\omega) = \int_0^\infty C(T) \cos \omega T \, dT = \pi m_I^2 \delta(\omega) + \frac{\sigma_I^2 R}{\omega^2 + R^2},$$ (8.25)

which is Lorentzian with half-width $\omega = R$ except for a zero frequency component.

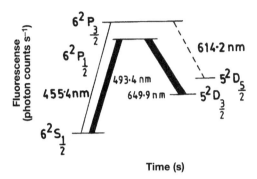

Fig. 8.4. Energy-level structure of Ba$^+$. The shelf level is the $5^2D_{5/2}$ state. The laser excitation is shown by the bold lines; the lamp excitation is indicated by the light solid line while the subsequent decay into the shelf level is indicated by the dashed line (after Ref. [191]).

So far we have considered the case of intermittent fluorescence for incoherent excitations in which the spectral energy density of the exciting radiation is a slowly varying function of frequency across each atomic line. In actual experiments in which coherent laser excitation is used, a number of coherent effects such as Rabi oscillations and Autler–Townes splitting of the weak atomic line have to be taken into account. The simplicity of the theory of incoherent excitations derives from the fact that the weak transitions can be treated as rate processes for which probabilities for upward and downward transitions (R_+ and R_-) exist. This feature fortunately survives for coherent excitations of the strong transition. The only difference is that the transition rates R_+ and R_- are expressed in terms of coherent parameters, such as Rabi frequencies, instead of spectral energy densities and Einstein B coefficients. For further details, see Kimble *et al.* [198] and references therein.

The turning on and off of the strong resonance fluorescence can, of course, only be observed over intervals long compared to the life-time of the strong transition but short compared to that of the weak transition. This coarse-graining of time relative to the life-time of the strong transition, however, leads to a loss of essentially quantum features of this radiation such as photon anti-bunching. The random interruptions of the strong fluorescence only encode the weak-transition time scale, and behave like a random classical telegraphic signal. This passage from the quantum to the classical description of the process by time averaging is worth further theoretical analysis.

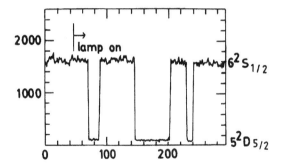

Fig. 8.5. A typical trace of the 493 nm fluorescence from the $6^2P_{1/2}$ level showing the quantum jumps after the hollow cathode lamp is turned on. The atom is definitely known to be in the shelf level during the fluorescence periods (after Ref. [191]).

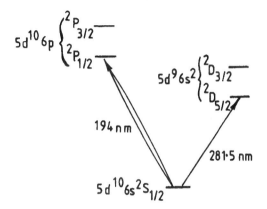

Fig. 8.6. Simplified optical energy-level diagram for HgII (after Ref. [193]).

8.3 Observation of quantum jumps

A number of experiments were carried out in 1986 in which quantum jumps were directly observed in individual laser-cooled ions. Nagourney *et al.* [191] observed the jumps between the $6^2S_{1/2}$ and $5^2D_{3/2}$ states of a single Ba$^+$ ion contained in a radio-frequency trap (Fig. 8.4). The cooling was done by two collinear laser beams which localized the ion to < 1 μm. The incoherent excitations to the $5^2D_{3/2}$ state caused the fluorescence from the $6^2P_{1/2}$ state to be interrupted for the > 30 s life-time of that state, after which the fluorescence reappeared (Fig. 8.5). A similar experiment was carried out by Sauter *et al.* [192]. Most of the jumps marked Raman–Stokes scattering from the $^2DS_{1/2}$ and $^2D_{3/2}$ levels. Bergquist *et al.* [193] used a laser-cooled HgII ion and observed the interruptions of the strong

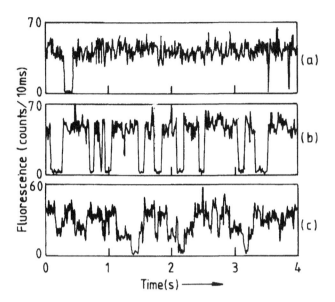

Fig. 8.7. Samples of the quantum-jump data. (a) The fluorescence counts detected with no resonance radiation exciting the weak $S_{1/2}$–$D_{1/2}$ transition. The few jumps observed were likely due to collisions with background Hg atoms and radiative decay to the $D_{3/2}$ state. (b) Both the 194 and 281.5 nm radiation are applied simultaneously. Compared to (a), the interruptions to the fluorescence signal are more frequent. (c) Two ions are trapped and cooled. The interruptions to the fluorescence signal show two levels corresponding to loss of fluorescence from one or both ions. The sampling rate for all the data is 10 ms per point; the length of each sample is 4 s (after Ref. [193]).

$5d^{10}6s^2S_{1/2} \rightarrow 5d^{10}6p^2P_{1/2}$ transitions (194 nm) due to the excitation of the metastable $5d^96s^2D_{5/2}$ level (281.5 nm) (Fig. 8.6). When the ion jumped back from the metastable D state to the ground state S, the S → P resonance fluorescence signal reappeared (Fig. 8.7). They also investigated the statistical properties of the quantum jumps and found evidence of photon anti-bunching in the emission from the D state.

It has been pointed out [200], however, that it is not necessary to assume collapse of the wave function in order to explain the interrupted fluorescence observed. The phenomenon can also be explained in terms of unitary Schrödinger evolution by taking into account the full state of the system, including the emitted photons.

9

Nonlocality

9.1 The Einstein–Podolsky–Rosen (EPR) argument

The most astonishing and counterintuitive aspect of quantum theory is the non-separability of two widely distant and non-interacting quantum systems that had interacted in the past. This became clear starting with the famous paper of Einstein, Podolsky and Rosen [201] and the subsequent papers of Bohr [202], Schrödinger [203], [204], Furry [205], Bohm [128], [129] and Bohm and Aharonov [206]. A detailed historical review will be found in Home and Selleri [207]. We will follow here Bohm's formulation which overcomes the unsatisfactory features of the original EPR formulation in which the wave function was not a stable solution of the time-dependent Schrödinger equation [208]. Another weakness of the original EPR wave function was that it was based on plane waves and did not describe two particles that were totally separated in space but rather two correlated particles present everywhere in space with the same probability.

Consider a pair (α, β) of spin-$\frac{1}{2}$ particles with the wave function

$$\Psi(x_1, x_2) = \eta_0 \, \Psi_\alpha(x_1) \, \Psi_\beta(x_2). \tag{9.1}$$

Here $\Psi_\alpha(x_1)$ and $\Psi_\beta(x_2)$ are the space parts of the wave functions of α and β respectively, and η_0 is the singlet state given by

$$\eta_0 = \frac{1}{\sqrt{2}} \left[u_\alpha(+)u_\beta(-) - u_\alpha(-)u_\beta(+) \right], \tag{9.2}$$

where $u_\alpha(+)$ and $u_\alpha(-)$ are eigenvectors corresponding to the eigenvalues $+1$ and -1 respectively of the Pauli matrix $\sigma_z(\alpha)$ representing the z-component of the spin of α, and $u_\beta(+)$ and $u_\beta(-)$ are the corresponding eigenvectors of the Pauli matrix $\sigma_z(\beta)$ for β. Note that the space part of the wave function (9.1) incorporates the time dependence while the spin part η_0 (9.2) is time independent and preserves its form over time.

141

We now suppose that $\Psi_\alpha(x_1)$ and $\Psi_\beta(x_2)$ are gaussian functions with moduli appreciably different from zero only in regions R_1 of width Δ_1 centred around x_1' and R_2 of width Δ_2 centred around x_2' respectively. A sufficient condition for separability of the particles α and β is

$$| x_2' - x_1' | \gg \Delta_1, \Delta_2. \tag{9.3}$$

If α and β move away from each other, the distance between the centres of the wave packets will increase linearly with time, and it can be shown that the Schrödinger equation will preserve this condition over time even though Δ_1 and Δ_2 increase with time. This allows one to assert that the two particles α and β are well localized in two small regions R_1 and R_2, well separated from each other so that all known physical interactions between them are negligible. In such a condition one would be inclined to infer that a measurement on one cannot in any way affect the other instantaneously. However, the presence of the factor η_0 in the wave function (9.1) leads to a puzzling non-local effect which is the essence of the so-called EPR paradox. Bohm's version of the EPR argument uses the following important properties of η_0:

(1) η_0 is not factorizable into a product of the wave functions for α and β.

(2) It is rotationally invariant.

(3) It predicts opposite results for the components of the spins of α and β measured along an arbitrary direction $\hat{n}$.

(4) It describes a state of α and β that has total squared spin zero.

Now, the EPR argument runs as follows. Consider a large set E of (α, β) pairs in the state (9.1). Let a measurement of $\sigma_z(\alpha)$ at time t_0 on all αs of a subset E_1 of E give the result $+1$ (or -1). Then a future measurement of $\sigma_z(\beta)$ at time $t(t > t_0)$ will certainly give the opposite result -1 (or $+1$). One can then invoke [201] the EPR *reality criterion*,

> If, without in any way disturbing a system, we can predict with certainty (i.e., with probability equal to unity) the value of a physical quantity, then there exists an element of physical reality corresponding to this physical quantity,...

to assign an element of reality λ_1 (or λ_2) to the βs of E_1, fixing a priori the result -1 (or $+1$) of the $\sigma_z(\beta)$ measurement.

If one further invokes [201] the Einstein *separability criterion*, namely

> ...since at the time of measurement the two systems no longer interact, no real change can take place in the second system in consequence of anything that may be done to the first system,...

then an element of reality λ_1 (or λ_2) cannot be instantaneously created by a *spooky* action-at-a-distance due to the measurement of $\sigma_z(\alpha)$. It follows therefore that λ_1 (or λ_2) must actually already belong to all βs of the entire ensemble E.

Now comes [201] the EPR criterion of *completeness*:

> ...every element of the physical reality must have a counterpart in the physical theory.

Accordingly, the particle β, having a predetermined value of $\sigma_z(\beta)$, must be assigned the eigenstate $u_\beta(-)$ (or $u_\beta(+)$) in the theory. This, together with property (3) applied to the z-axis, implies that the ensemble E must be described by a mixture of the factorizable state vectors $u_\alpha(+)u_\beta(-)$ and $u_\alpha(-)u_\beta(+)$ with the same weight *even before* the $\sigma_z(\alpha)$ measurement is made ($t < t_0$), which is incompatible with property (1). Therefore, either (a) the description of physical reality given by the wave function η_0 is incomplete, or (b) the EPR criterion of *local realism* (reality + separability) must be false.

EPR concluded their paper with the following remark:

> While we have thus shown that the wave function does not provide a complete description of the physical reality, we left open the question of whether or not such a description exists. We believe, however, that such a theory is possible.

9.2 Bell's theorem

9.2.1 Introduction

One possible way of completing quantum mechanics was to supplement it with additional *hidden* variables to restore causality and locality, though it is doubtful whether Einstein himself ever advocated such a programme. According to Jammer [209], Einstein never commended the use of hidden variables, but was rather in favour of 'a completely different theory which, perhaps, in approximation reduces to the existing quantum theory'. Bell [210], however, adopted the hidden variables option and showed that local hidden variable theories were incompatible with the statistical predictions of quantum mechanics, and could be experimentally falsified. He is thus credited with having brought down what had appeared as a purely philosophical issue to within the reach of empirical science, and thus establishing what has been called *experimental metaphysics*.

In his 1966 paper Bell [211] clarified the confusions that existed in the literature concerning hidden variable theories. In 1932 von Neumann [112] had enunciated his famous 'proof' that hidden variable models of quantum theory were impossible. Then Bohm's *causal interpretation* of quantum mechanics [129] involving hidden variables appeared in 1952, defying von Neumann's result and showing, in Bell's words, 'the impossible done'. In 1963 Jauch and Piron [212] produced a new version of von Neumann's impossibility proof. Meanwhile Gleason [213] had obtained a powerful mathematical theorem which strengthened von Neumann's claim! A relevant corollary to Gleason's theorem is that , if the dimensionality of the state space is greater than two, a less objectionable version of von Neummann's proof can be given. Bell provided a new and elementary proof of this corollary in which he essentially anticipated the conclusion of Kochen and Specker [214] that *non-contextual* hidden variable models of quantum systems associated with Hilbert spaces of three or more dimensions are impossible. (A hidden variable theory is *non-contextual* if the predictions for all quantum observables are independent of the measurement context.) Bell also constructed a specific model of spin-$\frac{1}{2}$ particles which provided a causal description of quantum mechanics in terms of hidden variables and which agreed with all the quantum mechanical predictions for spin measurements. This was in direct contradiction to von Neumann's theorem but consistent with the theorems of Gleason and of Kochen and Specker because the Hilbert space of spin-$\frac{1}{2}$ particles is two-dimensional. Bell showed precisely what went wrong with von Neumann's proof – his additivity assumption for expectation values, though true for quantum mechanical states, need not hold for the hypothetical dispersion-free states in hidden variable theories that reproduce the *measurable* quantum mechanical predictions *when averaged over*.

Having thus clarified the situation regarding hidden variable models of quantum theory, Bell analyzed Bohm's 1952 papers which he regarded as 'the most successful attempt' at 'interpolating some (preferably causal) space–time picture, between preparation of and measurement on states, that makes the quest for hidden variables interesting to the unsophisticated'. Bohm's model fell outside the remit of the Kochen–Specker result, involving as it does *contextual* hidden variables. However, Bell drew attention to the 'curious feature' in this model that the trajectory equations have a 'grossly nonlocal character', implying that 'in this theory an explicit causal mechanism exists whereby the disposition of one piece of apparatus affects the results obtained with a different piece. In fact, the Einstein–Podolsky–Rosen paradox is resolved in the way which Einstein would have liked least'. To the best of Bell's knowledge, however, there was 'no *proof* that *any* hidden variable account of quantum mechanics *must* have this extraordinary character'. He therefore concluded:

It would therefore be interesting, perhaps, to pursue some further 'impossibility proofs', replacing the arbitrary axioms ... by some condition of locality, or of separability of distant systems.

9.2.2 Proof of Bell's inequality

Consider a pair of spin-$\frac{1}{2}$ particles (α, β) formed somehow in the singlet spin state (9.2) and moving freely in opposite directions. When the spin component of one of these particles is measured in a certain direction, quantum mechanics *predicts* with 100% certainty that its partner's spin will be opposite *if* measured in the *same* direction. Such a correlation is called *perfect*. It is possible to construct deterministic, local realist models that can reproduce these perfect correlations, regardless of the direction in which the original measurement is made. Bell's achievement lay in realizing that this is not possible for more general measurements where the spin components of the two correlated particles are measured along *different* directions. The statistical correlations in such cases are called *imperfect*.

We will now sketch a simple proof of Bell's inequality for deterministic hidden variables. The same result also holds for stochastic hidden variables [215]. One can measure selected components of the spins $\vec{\sigma}(\alpha)$ and $\vec{\sigma}(\beta)$ using, say, Stern–Gerlach magnets. If measurement of $\vec{\sigma}(\alpha) . \vec{a}$, where $\vec{a}$ is some unit vector, gives the value $A(\vec{a}) = +1$, then according to (9.2) measurement of $\vec{\sigma}(\beta) . \vec{a}$ must yield the value $B(\vec{b}) = -1$, and vice versa. Thus, $A(\vec{a})$ and $B(\vec{b})$ are two-valued or dichotomic variables. This is a feature that will play an important role in the proof. Let λ denote a set of hidden variables required to fix the values of these observables A and B and complete the quantum mechanical description. The result A of measuring $\vec{\sigma}(\alpha) . \vec{a}$ is then determined by $\vec{a}$ and λ, and the result of measuring $\vec{\sigma}(\beta) . \vec{b}$ *at the same instant* is determined by $\vec{b}$ and λ, and

$$A(\vec{a}, \lambda) = \pm 1, \qquad (9.4)$$
$$B(\vec{b}, \lambda) = \pm 1. \qquad (9.5)$$

The crucial assumption is that the result B for particle β does not depend on the setting $\vec{a}$ of the magnet for particle α, nor A on $\vec{b}$. This is the Einstein separability assumption [1]:

> But on one supposition we should, in my opinion, absolutely hold fast: the real factual situation of the system S_2 is independent of what is done with the system S_1, which is spatially separated from the former.

Since the pair of particles is generally emitted by a source in a manner physically independent of the adjustable parameters $\vec{a}$ and $\vec{b}$, one assumes that the probability distribution $\rho(\lambda)$ characterizing the ensemble is independent of $\vec{a}$ and $\vec{b}$ and is normalized:

$$\int_\Gamma \rho(\lambda)\, d\lambda = 1. \tag{9.6}$$

Then the *correlation function* $P(\vec{a}, \vec{b})$ is the expectation value of the product of the two components $\vec{\sigma}(\alpha) \cdot \vec{a}$ and $\vec{\sigma}(\beta) \cdot \vec{b}$:

$$P(\vec{a}, \vec{b}) = \int \rho(\lambda)\, d\lambda\, A(\vec{a}, \lambda)\, B(\vec{b}, \lambda). \tag{9.7}$$

This should equal the quantum mechanical expectation value, which for the singlet state (9.1) is

$$\langle \vec{\sigma}(\alpha) \cdot \vec{a} \otimes \vec{\sigma}(\beta) \cdot \vec{b} \rangle = -\vec{a} \cdot \vec{b}. \tag{9.8}$$

The surprising result that Bell obtained was that this quantum mechanical result was incompatible with local realism. Define the quantity

$$\Delta = | P(\vec{a}, \vec{b}) - P(\vec{a}, \vec{b}') | + | P(\vec{a}', \vec{b}) + P(\vec{a}', \vec{b}') |. \tag{9.9}$$

It is a simple exercise to show that

$$| P(\vec{a}, \vec{b}) - P(\vec{a}, \vec{b}') | \le \int d\lambda \, | B(\vec{b}, \lambda) - B(\vec{b}', \lambda) |, \tag{9.10}$$

since $| A(\vec{a}, \lambda) | = 1$, and also that

$$| P(\vec{a}', \vec{b}) + P(\vec{a}', \vec{b}') | \le \int d\lambda \, | B(\vec{b}, \lambda) + B(\vec{b}', \lambda) |. \tag{9.11}$$

Adding (9.10) and (9.11) and using the result

$$| B(\vec{b}, \lambda) - B(\vec{b}', \lambda) | + | B(\vec{b}, \lambda) + B(\vec{b}', \lambda) | = 2, \tag{9.12}$$

because

$$| B(\vec{b}, \lambda) | = | B(\vec{b}', \lambda) | = 1, \tag{9.13}$$

one obtains Bell's inequality

$$\Delta \le 2. \tag{9.14}$$

Now consider a setting of the magnets such that they can be obtained by clockwise rotation of $\pi/4$ in the order $\vec{a}$, $\vec{b}$, $\vec{a}'$, $\vec{b}'$. It is then easy to see that $\Delta = 2\sqrt{2}$ if use is made of the quantum mechanical result (9.8). One can also show that this is the maximum value of Δ for all conceivable orientations of the magnets. But this is incompatible with Bell's inequality (9.14) which predicts a maximum value of 2 for Δ. Thus, wrote Bell [210],

> ...the statistical predictions of quantum mechanics are incompatible with separable predetermination.

In a theory in which parameters are added to quantum mechanics to determine the results of individual measurements, without changing the statistical predictions, there must be a mechanism whereby the setting of one measuring device can influence the reading of another instrument, however remote.

Since photons can be in either of two polarization states, the quantum mechanical treatment of photon polarization is similar to that of spin-$\frac{1}{2}$ particles, and the above proof of Bell's inequality for spin measurements can be easily adapted for polarization measurements of correlated photons in a zero angular momentum state:

$$| 0^{+} \rangle = \frac{1}{\sqrt{2}} [| R_\alpha \rangle | R_\beta \rangle + | L_\alpha \rangle | L_\beta \rangle] \tag{9.15}$$

$$= \frac{1}{\sqrt{2}} [| x_\alpha \rangle | x_\beta \rangle + | y_\alpha \rangle | y_\beta \rangle] \tag{9.16}$$

where $| R \rangle$ and $| L \rangle$ are single-photon states with right and left circular polarization, and $| x \rangle$ and $| y \rangle$ are single-photon states with linear polarization along the x- and y-axes respectively. Notice that the expression (9.16)) is invariant under rotations of the x- and y-axes.

9.2.3 The Clauser–Horne–Shimony–Holt (CHSH) inequalities

Although Bell's inequality points to a decisive experimental test of the entire family of local hidden variable theories, it is not in a form in which it can be directly tested in a realizable experiment because of three factors:

(a) Existing two-particle sources and/or analyzer–detector apparatuses are not efficient enough. For example, with detector efficiencies around 10–20 per cent it is impossible to be sure that a photon has been absorbed ($A(a)$ or $B(b)$ equals -1) rather than missed. Only a joint probability of double transmission *and* double detection can be measured.

(b) Bell assumes that it is possible to have perfect correlations for some directions b and b', whereas in practice there could be small departures from this ideal condition.

(c) Some additional assumptions concerning detector efficiencies that are plausible but untestable appear necessary to demonstrate the incompatibility of Bell's inequalities with quantum mechanics in realizable experiments.

Clauser, Horne, Shimony and Holt [216] (henceforth referred to as CHSH) generalized Bell's work to incorporate the above features and

arrive at new inequalities that are *stronger* than Bell's and that can be
used to test the entire family of local hidden variable theories.

Consider an ensemble of particles (which could be photons) moving so
that one enters apparatus $\mathscr{A}_a$ and the other apparatus $\mathscr{A}_b$, where a and
b are adjustable apparatus parameters. In each apparatus a particle must
select one of two channels labelled $+1$ and -1. Let the results of these
selections be represented by the dichotomic variables $A(a)$ and $B(b)$, each
of which can be ± 1 depending on whether the first or the second channel
is selected. Defining the statistical correlation function

$$P(a, b) = \int_{\Gamma} A(a, \lambda)\, B(b, \lambda)\, \rho(\lambda)\, \mathrm{d}\lambda \qquad (9.17)$$

where Γ is the total λ space, one has

$$
\begin{aligned}
| P(a, b) - P(a, c) | \;&\le\; \int_{\Gamma} | A(a, \lambda)B(b, \lambda) - A(a, \lambda)B(c, \lambda) | \, \rho(\lambda)\mathrm{d}\lambda \\
&= \int_{\Gamma} | A(a, \lambda)B(b, \lambda) | \, [1 - B(b, \lambda)B(c, \lambda)] \, \rho(\lambda)\mathrm{d}\lambda \\
&= \int_{\Gamma} [1 - B(b, \lambda)B(c, \lambda)] \, \rho(\lambda)\mathrm{d}\lambda \\
&= 1 - \int_{\Gamma} B(b, \lambda)\, B(c, \lambda)\, \rho(\lambda)\, \mathrm{d}\lambda. \qquad (9.18)
\end{aligned}
$$

Suppose that for some b' and b, $P(b', b) = 1 - \delta$ with $0 \le \delta \le 1$, i.e., the
correlation is not perfect ($\delta \ne 0$). In many experiments δ will be close but
not equal to zero. One can divide Γ into two regions Γ_+ and Γ_- such that
$\Gamma_{\pm} = \{\lambda \mid A(b', \lambda) = \pm B(b, \lambda)\}$. Then $\int_{\Gamma_-} \rho(\lambda)\mathrm{d}\lambda = \tfrac{1}{2}\delta$. Therefore,

$$
\begin{aligned}
\int_{\Gamma} B(b, \lambda)\, B(c, \lambda)\, \rho(\lambda)\, \mathrm{d}\lambda \;&=\; \int_{\Gamma} A(b', \lambda)\, B(c, \lambda)\, \rho(\lambda)\, \mathrm{d}\lambda \\
&\quad - 2 \int_{\Gamma_-} A(b', \lambda)\, B(c, \lambda)\, \rho(\lambda)\mathrm{d}\lambda \\
&\ge P(b', c) - 2 \int_{\Gamma_-} | A(b', \lambda)B(c, \lambda) | \, \rho(\lambda)\mathrm{d}\lambda \\
&= P(b', c) - \delta, \qquad (9.19)
\end{aligned}
$$

and so

$$| P(a, b) - P(a, c) | \le 2 - P(b', b) - P(b', c). \qquad (9.20)$$

In some experiments $P(a, b)$ depends only on the parameter difference
$b - a$. Defining $\alpha \equiv b - a$, $\beta \equiv c - b$ and $\gamma \equiv b - b'$, one gets

$$| P(\alpha) - P(\alpha + \beta) | \le 2 - P(\gamma) - P(\beta + \gamma). \qquad (9.21)$$

Unfortunately, because of the inefficiencies of present-day photodetectors,
these inequalities (9.20) and (9.21) cannot be tested using optical photons.

It is therefore convenient to interpret $A(a) = \pm1$ and $B(b) = \pm1$ to mean emergence or non-emergence of the photons from the respective filters which will be assumed to be linear polarization filters, and a and b will represent their orientations. Now introduce the four probabilities $w(a_\pm, b_\pm)$ where, for example, $w(a_+, b_-)$ is the probability that $A(a) = +1$ and $B(b) = -1$. These probabilities must satisfy the condition

$$w(a_+, b_+) + w(a_+, b_-) + w(a_-, b_+) + w(a_-, b_-) = 1. \tag{9.22}$$

It is also convenient to introduce an exceptional value ∞ of the parameters a and b to represent the removal of the corresponding polarizer. Then the following relations must hold when one of the polarizers is removed:

$$w(a_+, b_+) + w(a_+, b_-) = w(a_+, \infty), \tag{9.23}$$
$$w(a_+, b_+) + w(a_-, b_+) = w(\infty, b_+). \tag{9.24}$$

When both polarizers are removed, both the photons will certainly emerge, so that

$$w(\infty, \infty) = 1. \tag{9.25}$$

Now, the correlation function can be expressed as

$$P(a, b) = w(a_+, b_+) - w(a_+, b_-) - w(a_-, b_+) + w(a_-, b_-). \tag{9.26}$$

Introducing the four relations (9.22–9.25) this can be expressed as

$$P(a, b) = 4w(a_+, b_+) - 2w(a_+, \infty) - 2w(\infty, b_+) + 1. \tag{9.27}$$

Notice that only cases of double emission occur in this expression. Nevertheless, since one can measure only a joint probability for double emission *and* double detection, CHSH made an additional *ad hoc* assumption concerning the emission/detection process:

> ... when a pair of photons emerge from two regions of space where two polarizers are located, the probability of their joint detection (D_0) in two photomultipliers is independent of the presence and the orientation of the polarizers.

Denoting the joint probability for emission *and* detection by Ω, one obtains from this assumption

$$\begin{aligned}
\Omega(a, b) &= D_0 \, w(a_+, b_+), \\
\Omega(\infty, b) &= D_0 \, w(\infty, b_+), \\
\Omega(a, \infty) &= D_0 \, w(a_+, \infty), \\
\Omega(\infty, \infty) &= D_0,
\end{aligned} \tag{9.28}$$

where D_0 has been assumed in all cases to be the same. The rates for double detection are proportional to the number N_0 of photon pairs entering the solid angles defined by the optical apparatuses per second. Denoting by $R(a, b)$ the detection rate (the number of photon pairs detected per second), one has

$$R(a, b) = N_0 \Omega(a, b), \qquad (9.29)$$

$$R(\infty, b) = N_0 \Omega(\infty, b), \qquad (9.30)$$

$$R(a, \infty) = N_0 \Omega(a, \infty), \qquad (9.31)$$

$$R_0 = N_0 D_0, \qquad (9.32)$$

where $R_0 = R(\infty, \infty)$. Using (9.28) and (9.32) in (9.27), one obtains

$$P(a, b) = \frac{4R(a, b)}{R_0} - \frac{2R(a, \infty)}{R_0} - \frac{2R(\infty, b)}{R_0} + 1. \qquad (9.33)$$

The additional CHSH assumption has enabled the correlation function to be expressed only in terms of the coincidence rates which are measurable! One can now express the inequalities (9.20) and (9.21) for imperfect correlations in terms of experimental quantities. If $R(a, \infty)$ and $R(\infty, b)$ are found experimentally to be constants R_1 and R_2, the result is

$$| R(a, +b) - R(a, c) | + R(b', b) + R(b', c) - R_1 - R_2 \leq 0. \qquad (9.34)$$

In the special case in which $P(a, b) = P(a - b)$, this becomes

$$| R(\alpha) - R(\alpha + \beta) | + R(\gamma) + R(\beta + \gamma) - R_1 - R_2 \leq 0. \qquad (9.35)$$

Since Bell's inequality can be written in the form

$$-2 \leq P(a, b) - P(a, b') + P(a', b) + P(a', b') \leq +2, \qquad (9.36)$$

using expressions of the type (9.33) for the four correlation functions, it can also be written in the form

$$-1 \leq S \leq 0, \qquad (9.37)$$

where

$$S = \frac{R(a, b)}{R_0} - \frac{R(a, b')}{R_0} + \frac{R(a', b)}{R_0} + \frac{R(a', b')}{R_0} - \frac{R(a', \infty)}{R_0} - \frac{R(\infty, b)}{R_0}. \qquad (9.38)$$

Should every R function depend only on the relative angle between the polarizer axes, and these axes are so chosen that

$$a - b = a' - b = a' - b' = \phi \qquad (9.39)$$

$$a - b' = 3\phi, \qquad (9.40)$$

one obtains

$$-1 \leq \frac{3R(\phi)}{R_0} - \frac{R(3\phi)}{R_0} - \frac{R_1 + R_2}{R_0} \leq 0. \qquad (9.41)$$

For $\phi = \pi/8$ and $\phi = 3\pi/8$ this reduces to the Freedman inequality

$$\delta \equiv | \frac{R(3\pi/8)}{R_0} - \frac{R(\pi/8)}{R_0} | \leq \frac{1}{4} \tag{9.42}$$

which does not contain R_1 and R_2.

9.2.4 The Clauser–Horne (CH) assumptions

Clauser and Horne [217] extended these discussions in some respects. They characterized a broad class of theories and called them 'objective local theories' (OLT) and discussed the fundamental premises which motivate them.

They defined OLTs as theories in which the state of a pair (α, β) of correlated particles is characterized by a variable λ. If $p_\alpha(\lambda, a)$ $(p_\beta(\lambda, b))$ is the probability that the object $\alpha(\beta)$ in the state λ crosses the polarizer with parameter a (b) and is subsequently detected, and $p_{\alpha,\beta}(\lambda, a, b)$ is the probability of coincident counts from the two detectors, then OLTs are characterized by the *factored form*

$$p_{\alpha,\beta}(\lambda, a, b) = p_\alpha(\lambda, a)\, p_\beta(\lambda, b) \tag{9.43}$$

which expresses the locality condition for a pair of objective, well-localized systems between which there is no action-at-a-distance. The corresponding quantum mechanical probability does not in general admit such factorization.

The ensemble probabilities are written as weighted averages of the individual probabilities:

$$p_\alpha(a) = \int d\lambda\, \rho(\lambda)\, p_\alpha(\lambda, a),$$

$$p_\beta(b) = \int d\lambda\, \rho(\lambda)\, p_\beta(\lambda, b),$$

$$\Omega(a, b) = \int d\lambda\, \rho(\lambda)\, p_{\alpha,\beta}(\lambda, a, b). \tag{9.44}$$

In order to deduce inequalities, Clauser and Horne used the following mathematical result. Given six real numbers x, x', X, y, y', Y such that

$$0 \leq x, x' \leq X, 0 \leq y, y' \leq Y, \tag{9.45}$$

one must always have

$$- XY \leq xy - xy' + x'y + x'y' - x'Y - Xy \leq 0. \tag{9.46}$$

By taking

$$x = p_\alpha(a, \lambda), \quad y = p_\beta(b, \lambda),$$
$$x' = p_\alpha(a', \lambda), \quad y' = p_\beta(b', \lambda) \tag{9.47}$$

in (9.46), multiplying by $\rho(\lambda)$ and integrating over λ, and taking $X = Y = 1$ since the maximum value of these probabilities is unity, one obtains

$$-1 \leq \Omega(a,b) - \Omega(a,b') + \Omega(a',b) + \Omega(a',b') - p_\alpha(a') - p_\beta(b) \leq 0. \quad (9.48)$$

This is a *weak* inequality and is also *inhomogeneous* as it involves both single and double detection probabilities.

Clauser and Horne [217] showed that the incompatibility of this class of theories with quantum mechanics could not be tested with available analyzer–detector systems of low effficiencies unless an additional assumption was made. They stated an assumption which was *weaker* than the CHSH assumption and from which all the CHSH inequalities can be derived. The assumption was that

> ...for every emission of a photon in a state λ, the probability of a count with a polarizer in place is less than or equal to the probability with the polarizer removed. (*no enhancement hypothesis*).

With such an assumption the weak and inhomogeneous inequality can be converted into a *strong* and *homogeneous* inequality involving only double detection probabilities. The CH assumption implies that

$$p_\alpha(\lambda, a) \leq p_\alpha(\lambda, \infty)$$
$$p_\alpha(\lambda, a') \leq p_\alpha(\lambda, \infty),$$
$$p_\beta(\lambda, b) \leq p_\beta(\lambda, \infty),$$
$$p_\beta(\lambda, b') \leq p_\beta(\lambda, \infty). \quad (9.49)$$

Taking $X = p_\alpha(\lambda, \infty)$ and $Y = p_\beta(\lambda, \infty)$ and D_0 to be the double detection probability, one obtains from (9.4) and (9.5)

$$-D_0 \leq \Omega(a,b) - \Omega(a,b') + \Omega(a',b) + \Omega(a',b') - \Omega(a',\infty) - \Omega(\infty,b) \leq 0. \quad (9.50)$$

For a critical assessment of OLTs see [207].

9.2.5   Weak and strong inequalities

Defining

$$\Gamma = \Omega(a,b) - \Omega(a,b') + \Omega(a',b) + \Omega(a',b') \quad (9.51)$$

the inequalities (9.48) and (9.50) can be rewritten respectively as

$$-1 + p_\alpha(a') + p_\beta(b) \leq \Gamma \leq p_\alpha(a') + p_\beta(b) \quad (weak) \quad (9.52)$$

and

$$-D_0 + \Omega(a',\infty) + \Omega(\infty,b) \leq \Gamma \leq \Omega(a',\infty) + \Omega(\infty,b) \quad (strong) \quad (9.53)$$

These inequalities should be compared with the quantum mechanical predictions

$$p_\alpha(a') = \frac{\epsilon_+^1 \eta_1}{2}, \tag{9.54}$$

$$p_\beta(b) = \frac{\epsilon_+^2 \eta_2}{2}, \tag{9.55}$$

$$D_0 = \eta_1 \eta_2, \tag{9.56}$$

$$\Omega(a', \infty) = \frac{\epsilon_+^1 \eta_1 \eta_2}{2}, \tag{9.57}$$

$$\Omega(\infty, b) = \frac{\epsilon_+^2 \eta_1 \eta_2}{2}, \tag{9.58}$$

$$\Omega(a, b) = \frac{1}{4} \left[\epsilon_+^1 \epsilon_+^2 + \epsilon_-^1 \epsilon_-^2 \cos 2(a - b) \right] \eta_1 \eta_2 \tag{9.59}$$

where $\epsilon_\pm^1$ and $\epsilon_\pm^2$ are well-known parameters related to the transmittances of the two polarizers, while η_1 and η_2 are the quantum efficiencies of the two photodetectors. By substituting this into (9.51) one can show [235] that quantum mechanics predicts

$$\Gamma_{\max} = \frac{1}{2} \left[\epsilon_+^1 \epsilon_+^2 + \epsilon_-^1 \epsilon_-^2 \cos 2(a - b) \right] \eta_1 \eta_2, \tag{9.60}$$

$$\Gamma_{\min} = \frac{1}{2} \left[\epsilon_+^1 \epsilon_+^2 - \epsilon_-^1 \epsilon_-^2 \sqrt{2} \right] \eta_1 \eta_2. \tag{9.61}$$

Typical values of the experimental parameters being $\epsilon_+^1 = \epsilon_+^2 = 1, \epsilon_-^1 = \epsilon_-^2 = 0.95, \eta_1 = \eta_2 = 0.1$, one obtains

$$\Gamma_{\min} = -0.00138 \tag{9.62}$$

and

$$\Gamma_{\max} = 0.01138. \tag{9.63}$$

However, (9.52) and (9.53), which are predictions of OLTs, become respectively

$$-0.9 \leq \Gamma \leq 0.1 \quad (weak), \tag{9.64}$$

$$0 \leq \Gamma \leq 0.01 \quad (strong). \tag{9.65}$$

It is clear that the role of the supplementary CH assumption is to reduce the range of possible values of Γ allowed by local realism from 1 to about 0.01, *a reduction by two orders of magnitude.* Consequently, *the weak inequality is compatible with the quantum mechanical predictions (9.62) and (9.63) and therefore with the experimental results, but the strong inequality is violated.*

9.3   Experimental tests of Bell's inequality

9.3.1   Quantum optical tests

A comprehensive survey of a large number of experiments involving
photon pairs will be found in Clauser and Shimony [218], Pipkin [219],
and Home and Selleri [207]. The type of experimental arrangement that
is required in order to satisfy the Einstein–Bell locality condition has been
a matter of debate. For example, what should be the separation between
the two detectors so that the joint detection events can be regarded as
being space-like separated? There is the possibility , as pointed out by
Bell [210], that

> ...the settings of the instruments are made sufficiently in ad-
> vance to allow them to reach some mutual rapport by exchange
> of signals with velocity less than or equal to that of light. In
> that connection, experiments of the type proposed by Bohm
> and Aharonov [206], in which the settings are changed during
> the flight of the particles, are crucial.

The Aspect–Dalibard–Roger (ADR) experiment   The first experiment of
this type using variable polarizers was performed by Aspect, Dalibard
and Roger [220]. All previous experiments had used static setups in which
polarizers were held fixed for the whole duration of a run. Each static
polarizer was replaced by a setup involving a switching device followed
by two polarizers in two different orientations, $\vec{a}$ and $\vec{a}'$ on one side and
$\vec{b}$ and $\vec{b}'$ on the other (Fig. 9.1). Such an optical switch is able rapidly
to redirect the incident light from one polarizer to the other. The optical
switching was achieved by a Bragg reflection from an ultrasonic standing
wave in water. Light is incident at the Bragg angle $\theta_B = 5$ mrad (Fig. 9.2).
It is fully transmitted straight when the amplitude of the standing wave
is zero, which occurs twice during an acoustical period. A quarter of a
period later the amplitude of the standing wave is maximum and, for a
suitable value of the optical power (1 W), the light is then almost fully
deflected by $2\theta_B$. Switching between the channels occurs about every 10 ns.
Since this delay, as well as the life-time (5 ns) of the intermediate level of
a $(J = 0) \rightarrow (J = 1) \rightarrow (J = 0)$ cascade in calcium used to generate the
correlated photon pairs, is small compared to the time taken by light to
travel the distance $(L = 12$ m) between the two switches $(L/c \simeq 40$ ns), a
detection event on one side and the corresponding change of orientation
on the other side are separated by a space-like interval.

The coincidence rates found by ADR were only a few per second, with
an accidental background of about one per second. With S defined by

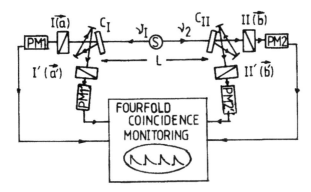

Fig. 9.1. ADR timing experiment with optical switches. Each switching device (C_I, C_{II}) is followed by two polarizers in different orientations. Each combination is equivalent to a polarizer switched fast between two orientations (after Ref. [220]).

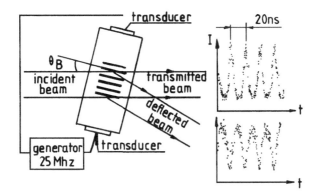

Fig. 9.2. Optical switch. The incident light is switched at a frequency around 50 MHz by diffraction at the Bragg angle on an ultrasonic standing wave (after Ref. [220]).

(9.38), they found $S = 0.101 \pm 0.020$ for $(\vec{a}, \vec{b}) = (\vec{b}, \vec{a}') = (\vec{a}', \vec{b}') = 22.5°$; $(\vec{a}, \vec{b}') = 67.5°$, in clear violation of (9.37) by five standard deviations, but in agreement with the quantum mechanical prediction $S_{QM} = 0.112$.

Although the settings of the polarizers were changed at a rate greater than c/L, the random delayed-choice scheme was not truly achieved because the changes were not random but quasi-periodic. However, the switches on the two sides were driven by different generators at different frequencies, and it was *assumed* that they functioned in an uncorrelated way. That such a situation can hide a conceptual difficulty has, however, been pointed out by Zeilinger [221].

Franson [222] has also argued that the ADR experiment does not appear to rule out a class of theories in which the outcome of an event is not

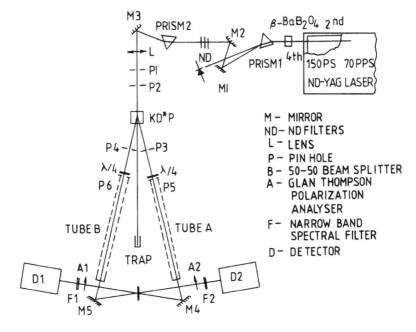

Fig. 9.3.   Schematic Shih–Alley experiment (after Ref. [223]).

determined until some time after its occurrence ('delayed determinism'). This class of theories includes not only quantum theory but also various local, realistic theories as well.

The Shih–Alley experiment   A random delayed-choice experiment is difficult to perform with photon pairs originating in atomic cascade emissions because of the necessity of a large solid angle of collection and the unknown time of emission of the photon pair. The possibility of a true delayed-choice experiment was however opened up by Shih and Alley [223] (see also Ou and Mandel [224]) using a different method of generating a correlated pair of photons at a definite time and with definite $\vec{K}$ vectors. They produced such pairs of 532 nm wavelength with the same linear polarization but slightly divergent directions of propagation by nonlinear parametric down-conversions in which phase-matching conditions have to be satisfied. Each photon was first converted to a definite polarization eigenstate by inserting a quarter- or half-wave plate into its path. It was then reflected by a turning mirror to superpose with the other at a beam splitter, producing two different types of EPR–Bohm eigenstates (Fig. 9.3).

To achieve clear space-like separation of the detection events the two detectors were separated by 50 cm and the source used was a 100 ps pulsed

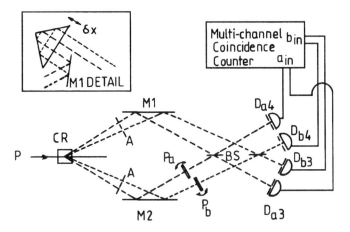

Fig. 9.4. Outline of the Rarity–Tapster apparatus (after Ref. [230]).

Nd-doped yttrium aluminum garnet laser. The 3 cm wave packet was sent through a 25 mm long deuterated potassium dihydrogen phosphate (KD*P) nonlinear crystal to produce the pair. Shih and Alley performed an explicit test of the Freedman inequality (9.42) and obtained $\delta = 0.34 \pm 0.03$, violating the inequality by three standard deviations. The result is, however, in good agreement with the quantum mechanical prediction $\frac{1}{4}\sqrt{2} \approx 0.35$.

A similar experiment has also been performed by Ou and Mandel [224]. It must be emphasized that these experiments were not in themselves truly random delayed-choice experiments, but opened up their possibility.

The Rarity–Tapster experiment   Violations of Bell's inequality need not be restricted to spins or polarizations [225], [226], [227] whose measurements depend on various auxiliary assumptions like the *no enhancement* hypothesis. Other tests based on fourth-order optical interference effects have been proposed [228], [229] and a violation of Bell's inequality based on phase and momentum, rather than spin, has been demonstrated [230]. Photon pairs of two different colours were selected by two double apertures placed to satisfy different phase-matching conditions at a down-conversion crystal. The different wavelengths were superposed at spatially separated points on two different beam splitters (Fig. 9.4) which could be arranged to be remote from each other (Fig. 9.5). On adjusting phase plates in the beams before the beam splitters, a nonlocal fourth-order interference effect was observed which violated Bell's inequality by several standard deviations.

Although the remote beams were differentiated by their colour in this experiment, this is by no means necessary, and an alternative arrangement

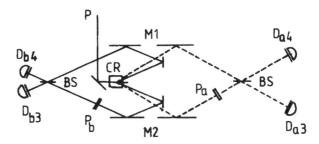

Fig. 9.5. Topologically equivalent apparatus to Fig. 9.4 showing the possibility of remote measurements on the separated colours (after Ref. [230]).

has been suggested by Franson [231] where each photon passes through a Mach–Zehnder interferometer.

The coincidence rates measured in this experiment were an order of magnitude higher than previous work due to the angular correlation of the parametrically down-converted photon pairs and the use of highly efficient solid state photon counting detectors in the near infra-red. In principle, detection efficiencies approaching 100% could be achieved in this type of experiment, and it may be possible to demonstrate a violation of Bell's inequality without the need for any auxiliary assumption.

9.3.2 *Particle physics tests*

Two possible ways of generating Bohm-type spin-correlated systems in particle physics are through the rare decays $\pi^0 \rightarrow e^+ + e^-$ and $\eta_c (2980) \rightarrow \Lambda + \bar{\Lambda}$. A different example involving hypercharge-correlated spin-0 particles, first suggested by Lee and Yang [232] and studied by various authors [233] [234], [235], [236] is the $K^0 \bar{K}^0$ system produced in the decay of the vector meson Φ. Assuming CP invariance, the state at the time of production ($t = 0$) can be written as

$$| \Psi_0 \rangle = \frac{1}{\sqrt{2}} \left[| K^0 \rangle_L | \bar{K}^0 \rangle_R - | \bar{K}^0 \rangle_L | K^0 \rangle_R \right] \tag{9.66}$$

$$= \frac{1}{\sqrt{2}} \left[| K_S \rangle_L | K_L \rangle_R - | K_L \rangle_L | K_S \rangle_R \right] \tag{9.67}$$

where the roman subscripts L and R indicate the left and right hemispheres, and

$$| K_L \rangle = \frac{1}{\sqrt{2}} \left[| K^0 \rangle + | \bar{K}^0 \rangle \right], \tag{9.68}$$

$$| K_S \rangle = \frac{1}{\sqrt{2}} \left[| K^0 \rangle - | \bar{K}^0 \rangle \right] \tag{9.69}$$

are the long-lived and short-lived neutral kaons which are odd and even eigenstates of CP respectively. If the right kaon is observed to be a K^0 (hypercharge $+1$) at some instant, then the state (9.67) predicts with certainty that the left kaon found at the same instant will be a $\bar{K}^0$ (hypercharge -1), and vice versa. On the other hand, if the right kaon is found to decay in the K_S mode (CP $= +1$), the state (9.67) predicts with certainty that the left kaon will be found to decay in the K_L mode (CP $= -1$) at some *future* time determined by the life-time of K_L.

Six [234] suggested that it should be possible to test this EPR-like situation by measuring the joint probability $\Omega_{-,-}(t_1, t_2)$ of detecting $\bar{K}^0$s on both the left and the right at times t_1 and t_2. Quantum mechanically this is given by

$$\Omega_{-,-}(t_1, t_2) = | \langle \bar{K}^0_L \, K^0_R \mid \Psi(t_1, t_2) \rangle |^2, \tag{9.70}$$

where

$$\begin{aligned} \mid \Psi(t_1, t_2) \rangle &= \frac{1}{\sqrt{2}} \left[\mid K_S \rangle_L \mid K_L \rangle_R \, e^{-i(\lambda_S t_1 + \lambda_L t_2)} \right] \\ &- \frac{1}{\sqrt{2}} \left[\mid K_L \rangle_L \mid K_S \rangle_R \, e^{-i(\lambda_L t_1 + \lambda_S t_2)} \right] \end{aligned} \tag{9.71}$$

is the state that evolves from $\mid \Psi_0 \rangle$, $\lambda_L = m_L - i\gamma_L/2$ and $\lambda_S = m_S - i\gamma_S/2$, $m_L(m_S)$ and $\gamma_L(\gamma_S)$ being the mass and decay width of $K_L(K_S)$ respectively. Hence

$$\begin{aligned} \Omega_{-,-}(t_1, t_2) &= \frac{1}{8} \left[e^{-(\gamma_S t_1 + \gamma_L t_2)} + e^{-(\gamma_L t_1 + \gamma_S t_2)} \right] \\ &- 2e^{-\gamma(t_1 + t_2)} \cos \Delta m (t_1 - t_2), \end{aligned} \tag{9.72}$$

where $\gamma = (\gamma_L + \gamma_S)/2$ and $\Delta m = m_L - m_S$.

Using a general argument based on the idea of Einstein locality or local realism, Selleri [235] derived an upper bound

$$\Omega_{-,-}(t_1, t_2) \leq \frac{1}{8} \left[e^{-(\gamma_S t_1 + \gamma_L t_2)} + e^{-(\gamma_L t_1 + \gamma_S t_2)} \right] \tag{9.73}$$

which is violated by the quantum mechanical prediction (9.72) whenever $\cos \Delta m (t_1 - t_2) < 0$. The maximum violation was calculated to be about 12 per cent for $\gamma_S(t_1 - t_2) \approx 5$.

The difficulty of testing such a prediction lies in the small life-time ($\approx 10^{-10}$ s) of K_L which implies that $t_1, t_2 \leq 10^{-10}$ s which makes the experimental uncertainties quite large. This may be circumvented by considering the time-integrated joint probabilities. This has been considered by Datta and Home [236] for the B^0–$\bar{B}^0$ system which is analogous to the K^0–$\bar{K}^0$ system. It should be possible to test these results in the Φ-factory at Frascati and the B-factories at SLAC (Stanford Linear Accelerator Center) and KEK (Japan).

9.4   The Greenberger–Horne–Zeilinger (GHZ) argument

Greenberger, Horne and Zeilinger [237] have formulated a new proof of
Bell's theorem (incompatibility of quantum mechanics and local realism)
for three (or more) well-separated spin-$\frac{1}{2}$ particles that are produced in a
decay *even in the case of perfect correlations* which lie at the core of the
EPR viewpoint but for which Bell's original theorem says nothing. Their
proof does not involve any inequality. Consider the state

$$| \Psi \rangle = \frac{1}{\sqrt{2}} [\, | \, 1, 1, 1 \rangle - | -1, -1, -1 \rangle \,] \tag{9.74}$$

which satisfies the conditions

$$O_i \, | \, \Psi \rangle = + | \, \Psi \rangle \tag{9.75}$$

with $i = 1, 2, 3$ for three hermitean operators

$$O_1 = \sigma_x^1 \sigma_y^2 \sigma_y^3; \; O_2 = \sigma_y^1 \sigma_x^2 \sigma_y^3; \; O_3 = \sigma_y^1 \sigma_y^2 \sigma_x^3. \tag{9.76}$$

It follows then that one can predict with certainty the result S_x of measuring the x component of spin of *any one* of the particles by distant
measurements of the y components of the spin of the other two. A similar result holds for the measurement S_y of the y component of its spin.
Applying the EPR·*reality criterion*, one can therefore regard the x and y
components of the spin of all the three particles to be elements of reality
having pre-assigned values ± 1. It follows from the EPR *separability* or
locality condition then that these values are independent of whichever
sets of spin measurements one might choose to make on these spatially
separated particles. We can therefore write the following relations:

$$S_x^1 S_y^2 S_y^3 = +1, \tag{9.77}$$
$$S_y^1 S_x^2 S_y^3 = +1, \tag{9.78}$$
$$S_y^1 S_y^2 S_x^3 = +1. \tag{9.79}$$

Since the measured values of the spin components are ± 1, it follows from
the above set of relations that

$$S_x^1 S_x^2 S_x^3 = +1. \tag{9.80}$$

However,

$$\sigma_x^1 \sigma_x^2 \sigma_x^3 \, | \, \Psi \rangle = - | \, \Psi \rangle, \tag{9.81}$$

which contradicts (9.80). This completes the GHZ proof.

   This powerful demonstration of a clear-cut contradiction has received
wide attention [238], [240], [239]. However, it remains to be seen if it can
be tested in a concrete physical situation (see section 2.1.5 of [207]).

9.5 Two-particle nonlocality *sans* inequalities

The GHZ argument uses three spin-$\frac{1}{2}$ particles, and therefore a Hilbert space of dimension six. The question that arises is: can nonlocality be demonstrated for two particles *without the use of inequalities*? The answer is yes, and the first proof was given by Heywood and Redhead [241]. However, they used two spin-1 particles, and therefore a Hilbert space of dimension six rather than the four required by Bell in his proof. It was Hardy [242] who first produced a proof without inequalities for two spin-$\frac{1}{2}$ particles in a Hilbert space of dimension four. We will sketch a simple argument given by Goldstein [243].

Consider a two-particle entangled state of the form

$$| \Psi \rangle = a | v_1 \rangle | v_2 \rangle + b | u_1 \rangle | v_2 \rangle + c | v_1 \rangle | u_2 \rangle \quad (abc \neq 0), \quad (9.82)$$

where $| u_i \rangle$ and $| v_i \rangle$ are a basis for particle $i, i = 1, 2$. Notice that the only term missing is $| u_1 \rangle | u_2 \rangle$. This is crucial for the proof. Let $U_i = | u_i \rangle \langle u_i |$ and $W_i = | w_i \rangle \langle w_i |$ with

$$| w_1 \rangle \quad = \quad \frac{1}{\sqrt{| a |^2 + | b |^2}} \left(a | v_1 \rangle + b | u_1 \rangle \right) \quad (9.83)$$

$$| w_2 \rangle \quad = \quad \frac{1}{\sqrt{| a |^2 + | c |^2}} \left(a | v_2 \rangle + c | u_2 \rangle \right) \quad (9.84)$$

be projection operators that represent physical observables. Each of these physical quantities can take the values 0 and 1 corresponding to the eigenvalues of U_i and W_i. It follows from (9.82) that the following results are predictions of quantum mechanics for the measurement of the variables U_i and W_i:

(1) $U_1 U_2 = 0$ because there is no $| u_1 \rangle | u_2 \rangle$ term,

(2) $U_1 = 0$ implies $W_2 = 1$,

(3) $U_2 = 0$ implies $W_1 = 1$, and

(4) $W_1 W_2 = 0$ with non-vanishing probability because $abc \neq 0$.

Let us now introduce the notion of realism by assuming that there exist some hidden variables λ which describe the state of each individual pair of particles. Assume that once the particles are produced in the state (9.82), they separate from each other sufficiently (so that there can be no appreciable interaction between them) and travel to two distant detectors where measurements of U_i and W_i can be made on them. Locality is ensured if the choice of a measurement on one side cannot influence the outcome of any measurement on the other side. Consider a run of the experiment for which U_1 and U_2 are measured and the result $U_1 U_2 = 0$ is obtained. That this will happen is guaranteed by (1). This can be realized either if (i) $U_1 = 0, U_2 \neq 0$, (ii) $U_2 = 0, U_1 \neq 0$ or (iii) $U_1 = U_2 = 0$. Suppose (i) holds. Then it follows from (2) that if W_2 had been measured on particle 2, one would have obtained the result $W_2 = 1$. If one assumes locality, then one can assert that, for this particular λ, one would have obtained the result $W_2 = 1$ even if W_1 had been measured on particle 1 rather than U_1, because what one chooses to measure on particle 1 cannot influence the outcome of any measurement on particle 2. Hence, for this run of the experiment, W_2 must be determined by the hidden variables to be equal to 1, i.e., $W_2(\lambda) = 1$. A similar argument holds for (ii) leading to $W_1(\lambda) = 1$ and for (iii) leading to $W_1(\lambda) = W_2(\lambda) = 1$. It therefore follows that W_1 and W_2 cannot simultaneously be zero for this run if local realism holds, which contradicts (4). It only remains to be shown that almost every spin state for the pair of spin-$\frac{1}{2}$ particles is of the form (9.82) for a suitable choice of basis. This can be done except for the maximally entangled state [243]. For a maximally entangled state, however, one has Bell's proof involving the famous inequality. Therefore, realistic interpretations of quantum mechanics must be nonlocal.

9.6   Signal locality

Although EPR correlations are nonlocal, there is a mathematical result in quantum mechanics, believed to be quite general, which shows that they cannot be used to send superluminal signals (at the statistical level). This may be called 'signal locality' as opposed to Einstein–Bell locality. As Pearle [244] puts it, 'It is surprising that nonrelativistic quantum theory does not allow superluminal communication to take place via correlated particles (e.g., EPR phenomena) when it can take place by other mechanisms (e.g., wave-packet travel or spread)'. The proof proceeds as follows. Let the initial state be a coherent superposition

$$| \Psi(t) \rangle = \sum_a c_a | a(t) \rangle \tag{9.85}$$

with $\langle \Psi(t) | \Psi(t) \rangle = 1$. The density operator $\rho(t)$ for such a 'pure' state, given by

$$
\begin{aligned}
\rho(t) &= | \Psi(t) \rangle \langle \Psi(t) | \\
&= U(t) | \Psi(0) \rangle \langle \Psi(0) | U^\dagger(t) \\
&= U(t) \rho(0) U^\dagger(t)
\end{aligned}
\tag{9.86}
$$

where $| \Psi(t)\rangle = U(t) | \Psi(0)\rangle$, has off-diagonal elements. The expectation value of any observable O in the state is given by

$$\langle O \rangle = \text{Tr } (\rho(t) O) = \text{Tr } \left(\rho(0) \tilde{O}(t) \right) \tag{9.87}$$

where $\tilde{O}(t) \equiv U^\dagger(t) O U(t)$. According to quantum theory, If a 'sender' makes a standard measurement on the state at some time t_s, the coherence of the state is completely destroyed, and the 'pure' state collapses completely into a 'mixed' state of components $| a(t_s)\rangle$ with probability coefficients $| c_a |^2$, with no quantum (EPR) correlations among them. The corresponding projection operators are $P_a(t_s) = | a(t_s)\rangle \langle a(t_s) |$, and the density operator is 'reduced' to the form

$$\hat{\rho}(t_s) = \sum_a P_a(t_s) \rho(t_s) P_a(t_s) \tag{9.88}$$

which is diagonal, the effect of collapse being to quench all the off-diagonal elements. If a 'receiver' measures the expectation value of O in such a mixed state at some future time $t_r > t_s$, the result will be

$$
\begin{aligned}
\langle O \rangle' &= \text{Tr } (\hat{\rho}(t_r) O) \\
&= \text{Tr } \left(U(t_r) \hat{\rho}(t_s) U^\dagger(t_r) O \right) \\
&= \text{Tr } \left(\hat{\rho}(t_s) \tilde{O}(t_r) \right) \\
&= \text{Tr } \sum_a \left(P_{a,}(t_s) \rho(t_s) P_a(t_s) \tilde{O}(t_r) \right) \\
&= \text{Tr } \sum_a \left(P_a(t_s) \rho(t_s) \tilde{O}(t_r) P_a(t_s) \right) \\
&= \text{Tr } \sum_a \left(P_a(t_s) \rho(t_s) \tilde{O}(t_r) \right) \\
&= \text{Tr } \left(\rho(t_s) \tilde{O}(t_r) \right) \\
&= \text{Tr } (\rho(t_r) O) \\
&= \langle O \rangle
\end{aligned}
\tag{9.89}
$$

where use has been made of two basic assumptions. One is

$$\left[\tilde{O}(t_r), P_a(t_s) \right] = 0 \tag{9.90}$$

whenever the interval between the acts of measurement by the 'sender' and the 'receiver' is space-like or the spatial separation between the subsystems (on which the 'sender' and 'receiver' act respectively) is so large that they are non-interacting and their Hilbert spaces are disjoint. This is the principle of micro-causality. The other assumption is

$$\sum_a P_a(t_s) = 1 , \tag{9.91}$$

which is the requirement of unitarity. This shows that as long as causality and unitarity hold, expectation values of observables cannot change as a result of state vector reduction or collapse, ensuring signal locality in spite of EPR-like nonlocal correlations.

Of course, the issue of superluminal signalling and its incompatibility with relativity can be meaningfully discussed only on the basis of a rigorous formulation of the EPR problem in relativistic quantum field theory. A proof of signal locality within the context of a relativistic quantum field theory has been given by Eberhard and Ross [245], and Ghose and Home [246] have shown how to represent an EPR state in a manifestly covariant fashion on a curved space-like surface, using the multiple-time Tomonaga–Schwinger formalism [247], [248]. This avoids the notion of a universal time, and clearly demarcates between the completion of the measuring process on a member of an EPR pair and its nonlocal effect on the state of its distant partner.

It should be pointed that all proofs of signal locality implicitly assume that measurements are *orthodox* or *ideal*. However, other kinds of *incomplete* measurements cannot be ruled out [249] which do not completely destroy the coherence of the original state. For such measurements the summation over the collapsed states occurring in (9.91) does not hold. An interesting and controversial example is the violation of signal locality in CP-violating kaon physics involving non-orthogonal states [250], [251], [252].

10

Tunneling times

10.1 Introduction

In classical physics it is meaningful to ask the question, 'How much time does a particle take to pass through a given region?' The interesting question in quantum mechanics is: does a particle take a definite time to tunnel through a classically forbidden region? The question has been debated ever since the idea that there was such a time in quantum theory was first put forward by MacColl way back in 1932 [253]. A plethora of times has since then been proposed, and the answer seems to depend on the interpretation of quantum mechanics one uses. A reliable answer is clearly of great importance for the design of high-frequency quantum devices, tunnelling phenomena (as, for example, in scanning tunneling microscopy), nuclear and chemical reactions and, of course, for purely conceptual reasons.

Most of the controversies centre around simple and intuitive notions in idealized one-dimensional models in a scattering configuration in which a particle (usually represented by a wave packet) is incident on a potential barrier localized in the interval $[a, b]$. Three kinds of time have been defined in this context. One, called the transmission time $\tau_T(a, b)$, is the average time spent within the barrier region by the particles that are *eventually* transmitted. Similarly, the reflection time $\tau_R(a, b)$ is the average time spent within the barrier region by the particles that are *eventually* reflected. These are *conditional* averages over mutually exclusive events. The third one, called dwell time $\tau_D(a, b)$, is, on the other hand, the *total* average time spent within the barrier region by *all* particles regardless of their ultimate fate. Unfortunately, there is no universally agreed procedure for calculating these times using the mathematical formalism of quantum mechanics. This is because time is a parameter or *c*-number and not a dynamical observable represented by a Hermitian operator τ, making the definition of an ideal measurement of time problematic. The only well-defined and well-established result is the dwell time.

A number of general properties or consistency checks have, however, been proposed for these times, based on intuitive physical requirements:

(1) *Additivity:* the time durations associated with physical processes occurring within the space interval $[a, b]$ must equal the sum of the corresponding times within the intervals $[a, c]$ and $[c, b]$ if $a \leq c \leq b$:

$$\tau_X(a, b) = \tau_X(a, c) + \tau_X(c, b) \qquad (10.1)$$

where X stands for T, R or D.

(2) *Mutual Exclusivity:* if the conditional averages exist, they must satisfy the identity

$$\tau_D = T \tau_T + R \tau_R, \qquad (10.2)$$

where T and R are respectively transmission and reflection probabilities satisfying the relation $T + R = 1$ (assuming that there is no absorption or loss within the barrier). This identity distinguishes tunneling in a scattering configuration from tunneling from a metastable state (such as alpha decay from an atomic nucleus). We will not discuss the latter process.

(3) *Reality:* the average duration of a process must obviously be real.

(4) *Positivity:* the duration of a physical process must also be non-negative.

A systematic approach has recently been developed by Brouard, Sala and Muga [254] (henceforth referred to as BSM) to define and classify transmission and reflection times, using the standard formalism of quantum mechanics, from which one can readily understand the root of the problem and gain a broad perspective covering a number of existing approaches. It follows from this approach that *none* of the times defined in this way satisfies all the conditions (1) to (4).

Several definitions of tunneling times have also been proposed based on incorporation of the concept of a trajectory into quantum mechanics, such as Bohm trajectories, Feynman paths, Wigner trajectories and wavepacket trajectories. We will first describe the BSM formalism and see how the many previously defined times, as well as new quantities, arise from it. We will then describe the trajectory approaches.

10.2   The Brouard–Sala–Muga (BSM) formalism

The root of the problem in standard quantum mechanics lies in the mathematical properties of the two projection operators corresponding to 'being in an interval $[a, b]$' and 'being transmitted' – they do not commute. Let us consider a static one-dimensional potential along the x-axis that is vanishingly small outside the domain $[x_1, x_2]$, not necessarily equal to $[a, b]$. Let a particle prepared in the state $|\psi(0)\rangle$ at $t = 0$ and $x = -\infty$ be incident on the potential from the left with a negligible negative momentum component. Then the probability to find the particle in $[a, b]$ at time t is

$$\langle D \rangle = \int_a^b |\langle x | \psi(t) \rangle|^2 \, dx, \tag{10.3}$$

where

$$D \equiv D(a, b) \equiv \int_a^b |x\rangle\langle x| \, dx \tag{10.4}$$

is the projector that selects the part of the wave function inside the interval $[a, b]$. Therefore, the average total time that the particle spends inside the interval $[a, b]$, called the 'dwell' or 'sojourn' time, is given by

$$\tau_D(a, b) \equiv \int_{-\infty}^{+\infty} \langle \psi(t) | D(a, b) | \psi(t) \rangle \, dt. \tag{10.5}$$

What about the average total time spent in the interval $[a, b]$ by a particle that is eventually *transmitted*? In order to answer this question one has to consider the projector

$$P \equiv \int_0^\infty |p^{(-)}\rangle\langle p^{(-)}| \, dp, \tag{10.6}$$

where the states $|p^{(-)}\rangle$ are eigenstates of the total Hamiltonian $H = H_0 + V$ and are asymptotic incoming solutions of the Lippmann–Schwinger equation:

$$|p^{(\pm)}\rangle = |p\rangle + \lim_{\epsilon \to +0} \frac{1}{E \pm i\epsilon - H} V |p\rangle. \tag{10.7}$$

When acting on an arbitrary square integrable state, P selects the part of the state that will have positive momentum at $t = +\infty$. Therefore, the probability of finding the particle transmitted in the infinite future (the transmittance) is $T = \langle P \rangle$. The complementary projector

$$Q \equiv \int_{-\infty}^0 |p^{(-)}\rangle\langle p^{(-)}| \, dp, \tag{10.8}$$

selects the part of the incident state that will be finally reflected, having negative momentum at $t = +\infty$. Therefore the reflectance R is given by $R = \langle Q \rangle = 1 - T$.

If each particle has a unique trajectory as in classical physics, then those trajectories corresponding to particles that are transmitted through the barrier are distinct from those that correspond to particles that are reflected by it. Consequently, the detection of a particle inside or outside the interval $[a, b]$ followed by an observation to check that it has been transmitted through the barrier is equivalent to performing these operations in the opposite order in time. But this is not true in quantum mechanics because the projectors D (10.4) and P (10.6) do not commute. This is basically the root of the problem – there is no unique way of constructing Hermitian operators in quantum mechanics from two non-commuting observables. In the present case the situation is compounded by the fact that D and P are projection operators, so that even the symmetrized product of P and D is not unique. Using $P + Q = 1$, $P^2 = P$ and $D^2 = D$, one can easily verify, for example, that

$$(PD)_{symm} = \frac{[P, D]_+}{2} = PDP + \frac{1}{2}(PDQ + QDP)$$

$$= PDP + \frac{1}{3}(PDQ + QDP).$$

There are therefore an infinite number of possible 'times' that one can define from suitable combinations of P and D. One can adopt one of three different attitudes in this situation :

(a) One can explore simple combinations of P and D that correspond to certain observables. This method not only systematically leads to the 'times' proposed earlier by various authors from intuitive notions but also to some new quantities.

(b) One can dismiss the question as meaningless and attempt to frame a new question that can be answered without invoking P and D simultaneously. This has also led to some interesting results concerning arrival times.

(c) One can regard the usual interpretation of quantum mechanics to be incomplete, and introduce a well-defined trajectory into quantum theory as, for example, in Bohm's causal interpretation.

The first two options lie within the mathematical framework of standard quantum mechanics and were explored by BSM. We will now examine these options one by one.

10.2.1 Combinations of P and D

The PDP resolution   If one *first* projects out the part of the state that will be transmitted in the future and *then* calculates the corresponding dwell time using D, one obtains

$$\tau_T^{PDP} = \frac{1}{T} \int_{-\infty}^{+\infty} \langle P\,\psi(t)\,|\,D\,|\,P\,\psi(t) \rangle \, dt. \tag{10.9}$$

Unfortunately, the physical interpretation of this time is obscured by the fact that $\langle x \mid P \psi(0) \rangle$ is in general non-zero *on both sides of the barrier* even when the initial packet $\langle x \mid \psi(0) \rangle$ is restricted to be entirely on the left. However, in some circumstances (such as in the case of spin precession in weak magnetic fields) it does acquire a physical significance.

One can also define a dwell time τ_T^{QDQ} by replacing P by Q and T by R in (10.9). An interesting relation between these two dwell times follows from the resolution

$$D = PDP + QDQ + PDQ + QDP$$

which can be proved by using $P + Q = 1$ twice. It follows from this that

$$\tau_D = T\tau_T^{PDP} + R\tau_R^{QDQ} + 2\mathrm{Re}\,[\tau_{\mathrm{int}}], \qquad (10.10)$$

where

$$\tau_{\mathrm{int}} = \int_{-\infty}^{+\infty} \langle \psi(t) \mid PDQ \mid \psi(t) \rangle \, dt \qquad (10.11)$$

is a complex quantity (because PDQ is non-Hermitian). The interference term is purely quantum mechanical in origin, and reflects the fundamental quantum mechanical feature that the labels 'to be transmitted' and 'to be reflected' cannot be attached to particles *before* any measurement is made when the state is a coherent linear superposition of both the probability amplitudes.

The DPD resolution One can alternatively *first* select the part of the wave function within the interval $[a, b]$ at time t by means of D, and *then* calculate the transmission probability using P. One then obtains

$$\tau_T^{DPD} = \frac{1}{T} \int_{-\infty}^{+\infty} \langle \psi(t) \mid DPD \mid \psi(t) \rangle \, dt. \qquad (10.12)$$

Since

$$D = DPD + DQD,$$

one obtains the relation

$$\tau_D = T\tau_T^{DPD} + R\tau_R^{DQD} \qquad (10.13)$$

without any interference term.

Complex times Complex times can also be defined through the resolution

$$D = PD + QD,$$

because PD and QD are non-Hermitean operators. One obtains

$$\tau_T^{PD} = \frac{1}{T} \int_{-\infty}^{+\infty} \langle \psi(t) \mid PD \mid \psi(t) \rangle \, dt \qquad (10.14)$$

for transmission and τ_R^{QD} for reflection. It follows from the resolution of D used in this case that $\tau_D = T\tau_T^{PD} + R\tau_R^{QD}$.

Using $1 = P + Q$ in (10.14), one establishes the relation

$$\tau_T^{PD} = \tau_T^{PDP} + \frac{1}{T}\tau_{int},\tag{10.15}$$

and analogously

$$\tau_R^{QD} = \tau_R^{QDQ} + \frac{1}{R}\tau_{int}^*.\tag{10.16}$$

The moduli of these complex times τ_T^{PD} and τ_R^{QD} are known in the literature as the Büttiker–Landauer times [255].

Hermitian and anti-Hermitian operators   One can easily establish the resolution

$$D = \frac{1}{2}[P,D]_+ + \frac{1}{2i}i[P,d]_- + \frac{1}{2}[Q,D]_+ + \frac{1}{2i}i[Q,D]_-,$$

from which it follows that

$$\tau_T^{[P,D]_+/2} = \frac{1}{T}\int_{-\infty}^{+\infty}\left\langle\psi(t)|\frac{[P,D]_+}{2}|\psi(t)\right\rangle dt,\tag{10.17}$$

$$\tau_T^{[P,D]_-/2i} = \frac{1}{T}\int_{-\infty}^{+\infty}\left\langle\psi(t)|\frac{[P,D]_-}{2i}|\psi(t)\right\rangle dt\tag{10.18}$$

are also transmission times; they are, respectively, the real and imaginary parts of τ_T^{PD}.

It is easy to check, using $D(a,b) = D(a,c) + D(c,b)$, that all times that can be written in terms of only one D operator (such as $\tau_D, \tau_X^Y, Y = PDP, PD, [P,D]_+/2, [P,D]_-/2i$) are additive, whereas τ_T^{DPD}, τ_R^{DQD} as well as the Büttiker–Landauer times $|\tau_T^{PD}|$ and $|\tau_R^{QD}|$, which are not linear in D, are not so.

All times defined in terms of the Hermitian operators D, PDP, DPD and $[P,D]_+/2$ are real. This leaves only τ_X^{PD} and τ_{int} which are complex.

All real times that are integrals of a probability density, such as the dwell time τ_D, τ_X^{PDP} and τ_X^{DPD} are non-negative. So are also, trivially, the Büttiker-Landauer times. The positivity of the real and imaginary parts of complex times such as τ_T^{PD} and τ_R^{QD} depends on the shape of the barrier.

It follows from the resolution $D = DPD + DQD$ that τ_T^Y ($Y = DPD$, PD and $[P,D]_+/2$) as well as τ_R^Y ($Y = DQD$, QD and $[Q,D]_+/2$) satisfy the mutual exclusivity requirement (10.2), whereas τ_T^{PDP} and τ_R^{QDQ} satisfy the relation (10.10) with an additional interference term. The imaginary parts of τ_T^{PD} and τ_R^{QD}, on the other hand, satisfy the relation

$$T\tau_T^{[P,D]_-/2i} + R\tau_R^{[Q,D]/2i} = 0.\tag{10.19}$$

Finally, the Büttiker–Landauer times $|\tau_T^{PD}|$ and $|\tau_R^{QD}|$ do not satisfy the requirement (10.2) when weighted by T and R respectively.

Thus, as we mentioned before, *none* of the times that can be defined with the help of the non-commuting operators D and P satisfies all the reasonable physical requirements (1) to (4).

Relationship to probability theory   The probabilities for the basic events 'being within $[a, b]$' and 'eventually passing the barrier' are defined by

$$p(D) \equiv \langle D \rangle \;=\; \langle \psi(t) | D | \psi(t) \rangle, \tag{10.20}$$

$$p(P) \equiv \langle P \rangle \;=\; \langle \psi(t) | P | \psi(t) \rangle = T, \tag{10.21}$$

and the conditional probabilities $p(D|P)$ and $p(P|D)$ by

$$p(D|P) \;=\; \frac{\langle \psi(t) | PDP | \psi(t) \rangle}{T}, \tag{10.22}$$

$$p(P|D) \;=\; \frac{\langle \psi(t) | DPD | \psi(t) \rangle}{\langle D \rangle}. \tag{10.23}$$

In terms of these probabilities the traversal times τ_T^{DPD} and τ_T^{PDP} can be expressed as

$$\tau_T^{DPD} \;=\; \frac{1}{p(P)} \int_{-\infty}^{+\infty} p(D)\, p(P|D)\, \mathrm{d}t, \tag{10.24}$$

$$\tau_T^{PDP} \;=\; \int_{-\infty}^{+\infty} p(D|P)\, \mathrm{d}t. \tag{10.25}$$

The BSM approach can also be used for non-normalizable plane wave states in the stationary regime, as well as for multi-channel scattering [254].

10.2.2 *Average passage or arrival times*

Let us now examine the second option, namely to accept that it is meaningless within quantum mechanics to ask the question about transmission time, forget about the operators D and P and look instead at the asymptotic properties of *fluxes* to see what can be learnt about the temporal aspects of tunneling from a hydrodynamic approach [254], [256] . Consider a wave packet impinging on the barrier from the left, but now take the interval $[a, b]$ sufficiently well outside the barrier $[x_1, x_2]$ so that the passage of the incident wave packet through the point a can be well separated from the subsequent passage of the reflected wave packet through the same point in the opposite direction so that there is no interference between them. Let t_c be an arbitrary instant of time in the interval between these two events during which the flux at the point a is zero. Then the dwell time is given by [256], [257]

$$\tau_D = \int_0^\infty [J(b,t') - J(a,t')]\, t'\, dt', \tag{10.26}$$

where $J(x,t) = (\hbar/m)\,\mathrm{Im}[\psi * (x,t)\partial\psi(x,t)/\partial x]$ is the current density at point x at time t. In the absence of the barrier one has

$$\int_0^\infty J(b,t')\, dt' = \int_0^\infty J(a,t')dt' = 1,$$

so that $J(x,t)$ can be interpreted as a distribution function for passing instants, and

$$\langle t \rangle = \int_0^\infty t' J(x,t')\, dt' \tag{10.27}$$

is the average instant of passage through the point x. Equation (10.26) can therefore be written as

$$\tau_D = \langle t \rangle_b - \langle t \rangle_a. \tag{10.28}$$

This shows that the dwell time can be regarded as the difference between the average outgoing and incoming instants.

Now consider the situation in the presence of the barrier. The current passes through a unaffected before t_c, and after some time, during which the flux is zero at a, the reflected part returns through a with a negative current. Therefore

$$\int_0^\infty J(b,t')\, dt' = T,$$
$$\int_0^{t_c} J(a,t')\, dt' = 1,$$
$$\int_{t_c}^\infty J(a,t')\, dt' = -R, \tag{10.29}$$

and we can write (10.26) as

$$\tau_D = T \langle t \rangle_b^{\mathrm{out}} - \langle t \rangle_a^{\mathrm{in}} + R \langle t \rangle_a^{\mathrm{out}}, \tag{10.30}$$

where

$$\langle t \rangle_b^{\mathrm{out}} = \frac{1}{T} \int_0^\infty J(b,t')\, t'\, dt', \tag{10.31}$$

$$\langle t \rangle_a^{\mathrm{in}} = \int_0^{t_c} J(a,t')\, t'\, dt', \tag{10.32}$$

$$\langle t \rangle_a^{\mathrm{out}} = -\frac{1}{R} \int_{t_c}^\infty J(a,t')\, t'\, dt'. \tag{10.33}$$

Using $T + R = 1$, one recovers (10.2) with

$$\tau_T = \langle t \rangle_b^{\mathrm{out}} - \langle t \rangle_a^{\mathrm{in}}, \tag{10.34}$$

$$\tau_R = \langle t \rangle_a^{\text{out}} - \langle t \rangle_a^{\text{in}}. \tag{10.35}$$

This has been made possible by choosing an instant t_c when the flux at a is zero, ensuring the absence of any interference between the incident and reflected wave packets. It corresponds to the asymptotic condition used in detailed calculations of stationary phase times which show that in the limit of a narrow wave packet peaked around a certain momentum, these transmission and reflection times tend to averages of the phase times over the initial momentum distribution [258] , [259], [254].

This separation of incident and reflected wave packets is obviously not possible for arbitrary a and b for which the conditions (10.29) break down. The formalism can, however, be generalized to this case also by separating the current density J into its positive and negative components [256] .

10.3 Trajectories and tunneling times

It should be clear by now that a 'tunneling time theory is on the edge of standard quantum mechanics and possibly extensions of the usual formalism and/or its interpretation are required' [256]. Let us therefore turn to the option of incorporating trajectories into quantum mechanics.

10.3.1 Bohm trajectories

The causal and ontological interpretation of de Broglie and Bohm [260], [23], [261] leads exactly to the same observational predictions as any of the conventional interpretations of quantum mechanics for every problem or question that is well posed in both interpretations. As we have seen, the transmission time τ_T of a particle through a barrier is not well defined in the conventional interpretations, whereas the concept is not only meaningful and precise in the Bohm interpretation, it is central to it. According to this interpretation, quantum mechanics is *completed* by introducing the position x of a particle as a so-called 'hidden variable' *in addition* to its wave function $\psi(x, t)$ that satisfies the usual Schrödinger equation. Thus, a particle is a *real* particle that is guided by its wave function so that at each instant of time it *has* (in the *ontological* sense) both a well-defined position $x = x(x_0, t)$ and velocity

$$v(x, t) = \frac{dx}{dt} = \left[\frac{J(x, t)}{\rho(x, t)} \right]_{x=x(x_0,t)}, \tag{10.36}$$

where x_0 is the particle's (unknown) initial position and ρ and J are the probability density and probability current density respectively. Equation (10.36) is called the 'guidance condition'. Since the initial positions are unknown, it is assumed that their distribution is given by quantum mechanics, namely by $\rho(x_0, 0)$. Since ρ satisfies the continuity equation

$$\frac{\partial \rho}{\partial t} + \operatorname{div} \vec{J} = 0, \tag{10.37}$$

once the actual distribution agrees with that predicted by quantum mechanics at time $t = 0$, the two are guaranteed to agree at all later times. Quantum mechanical expectation values are therefore replaced by averages over the ensemble of real particles. This is how observational agreement with standard quantum mechanics is built into this interpretation. Note that there is no conflict with the uncertainty relations in a particle *having* both a precise position and a precise velocity at each instant of time. This is because, as a result of the initial distribution being fixed by quantum mechanics, any *measurement* of the particle's position and velocity at a later time is guaranteed to produce a scatter in position and momentum values in agreement with the uncertainty relations. There is, however, no place for an 'uncertainty principle' in this interpretation – *the uncertainties are not inherent in the conceptual structure but arise due to 'unknown' initial conditions*. They are therefore of the same nature as in classical statistical mechanics. It is in this sense that the interpretation is *causal*.

In practice, the trajectories $x(t)$ are calculated as functions of the initial positions x_0 by solving the Schrödinger equation to calculate J and ρ and then integrating the guidance condition (10.36). Leavens [262] has used this formalism to calculate the mean dwell, transmission and reflection times and the distribution of arrival times for massive particles. If a particle is prepared in the state $\psi(x, t = 0)$, the mean dwell time $\tau_D(a, b)$ is the ensemble average time spent by the particle within the barrier region $[a, b]$ and is given by

$$\tau_D(a, b) = \int_0^\infty dt \int_a^b dx \, | \, \psi(x, t) \, |^2. \tag{10.38}$$

Now, an important feature of Bohm trajectories is that they do not cross or even touch one another in real place, i.e., if $x_0 \neq x_0'$, then $x(x_0, t) \neq x(x_0', t)$ for any t. This means that there is a bifurcation curve $x_c(t)$ separating transmitted from reflected trajectories, implicitly given by

$$T = \int_{x_c(t)}^\infty dx \, | \, \psi(x, t) \, |^2, \tag{10.39}$$

where T is the transmission probability. Hence, the mean time spent by a transmitted particle in the barrier region is given by the ensemble average

$$\tau_T(a, b) = \frac{1}{T} \int_0^\infty dt \int_a^b dx \, | \, \psi(x, t) \, |^2 \, \Theta(x - x_c(t)). \tag{10.40}$$

Similarly, the mean time spent by a reflected particle in the barrier region is given by

$$\tau_R(a, b) = \frac{1}{R} \int_0^\infty dt \int_a^b dx \, |\psi(x, t)|^2 \, \Theta(x_c(t) - x), \qquad (10.41)$$

where $R = 1 - T$ is the reflection probability. It follows therefore that all the conditions (1) to (4) are satisfied by these times for wave packets. However, the mutual exclusivity condition (2) (10.2) does not hold in the stationary limit where averaging over transmitted and reflected components is implicit [256], [263], [264].

Moreover, Leavens [265] has shown that the distribution of arrival times at the point $x = b$ is given by

$$P\,[t(b)] = \frac{1}{T} \, J\,[b, t(b)] \qquad (10.42)$$

when there are no re-entrant trajectories, and by

$$P\,[t(b)] = \frac{|\, J\,(b, t(b))\,|}{\int_0^\infty dt\, |\, J(b, t)\,|} \qquad (10.43)$$

when there are re-entrant trajectories.

Leavens [262] has also shown that the unique transmission and reflection times in the Bohm approach are *not* included in the infinite set of possibilities generated by the BSM approach. The basic reason is that the transmission–reflection decompositions are fundamentally different in the two approaches – particle-like,

$$
\begin{aligned}
|\psi|^2 &= \left[|\psi|^2\right]_T + \left[|\psi|^2\right]_R, \\
\left[|\psi|^2\right]_T &\equiv \Theta(x - x_c(t))|\psi|^2, \\
\left[|\psi|^2\right]_R &\equiv \Theta(x_c(t) - x)|\psi|^2,
\end{aligned}
\qquad (10.44)
$$

in the Bohm trajectory approach, and wave-like,

$$\psi = \psi_T + \psi_R, \qquad (10.45)$$

in the BSM projector approach. As Leavens puts it: 'The essential difference stems from the fact that Bohmian mechanics consistently treats a "particle" as a particle at all times not just, as in conventional quantum mechanics, at the instant(s) when an ideal measurement is made of its position' [266].

It is for this reason that tunneling times through a potential barrier (in a scattering configuration) is a situation in which Bohm's theory can make a definite prediction whereas conventional quantum mechanics can make none at all. Some of the theoretical and experimental prospects for a clear-cut comparison have been summarized by Cushing [267]. One has to bear in mind that if a measuring device is not included in the calculation of these average times, they must be regarded as intrinsic quantities and their measurability has to be thoroughly investigated. Leavens [262] has shown that quantum clocks augur reasonably well only for the measurability of the intrinsic average dwell time.

Further, according to the Bohmian school of thought, the trajectory concept can be extended to the relativistic domain for spin-$\frac{1}{2}$ fermions but not for spin-0 and spin-1 bosons. A causal and ontological interpretation of the quantized field theory of such bosons, however, exists in which the field coordinates $\psi(x)$ replace the particle positions x. Consequently, it is believed [23] that the photon does not have a trajectory, and so there is no Bohmian theory of tunneling times for photons. An alternative interpretation of relativistic bosons (massive and massless) as particles with Bohmian trajectories (below the threshold for particle creation and annihilation) has recently been developed, based on the Hamiltonian form of the Kemmer equation and the associated constraints [268], [43]. Tunneling time calculations for photons using this formalism are in progress.

10.3.2 *Feynman paths*

Sokolovski and coworkers [269], [270], [271] have applied the Feynman path integral techniques to non-standard functionals to derive expressions for τ_X. They propose a formal generalization of the classical time concept to the quantum domain. For one-dimensional motion the classical time spent by a particle in a region Ω is

$$\tau_{cl}^{\Omega} [x(t)] = \int_{t_i}^{t_f} dt \int_{\Omega} dx \, \delta [x - x(t)], \qquad (10.46)$$

where $x(t)$ is the classical path from the initial point $x_i(t_i)$ to the final point $x_f(t_f)$. A possible quantum generalization of this is the path-integral average

$$\tau^{\Omega}(x_i, t_i; x_f, t_f) = \langle \, \tau_{cl}^{\Omega} [x()] \, \rangle_{\text{paths}}, \qquad (10.47)$$

where $x()$ is an arbitrary path between the initial and final points. In general τ^{Ω} turns out to be complex. Moreover, Sokolovski and Baskin [269] pointed out that their prescription (10.47) for τ^{Ω} is not even unique. For instance, a number of possible alternatives would have the same classical limit τ_{cl}^{Ω}. For example, the right-hand side of (10.47) can be replaced by either its real part or its absolute value.

It is important to remember that there is a fundamental difference between Feynman paths and Bohm trajectories. In the Feynman approach a particle prepared at an initial point (x_i, t_i) reaches at a later time t all points x where the wave function is non-zero, and for every such point does so by following all conceivable paths joining it to the initial point.

None of these paths coincides with a Bohm trajectory unless $x = x(x_i, t)$, and in that case the one that does is highly atypical. Also, typical Feynman paths are highly irregular on a fine scale, and are non-differentiable [272]. Bohm trajectories, on the other hand, are differentiable because they are obtained by integrating the guidance condition (10.36). They are a set of actually realizable trajectories. Feynman paths are not. Had they been so, the Bohm and Feynman space–time pictures would have been completely incompatible. However, it is widely recognized that Feynman paths are to be interpreted as useful mathematical constructs and not actually realizable trajectories.

Since the derivations of expressions for τ_X involve a non-unique extension of the standard formalism, there is no reason to regard the calculated τ_X as a unique prediction of quantum mechanics for an ideal measurement. Such an ideal measurement may not even exist in principle in the absence of a corresponding Hermitian operator. In such cases it is important to include the measuring apparatus (a quantum clock) explicitly in the calculation in order to compare the calculated value with experiment. We will see when considering clocks that although the Sokolovski–Baskin procedure is formal and non-unique, the results can be given physical interpretation through connections to clearly defined model situations.

Iannaccone and Pellegrini [273] have developed a Feynman path-integral formulation similar to that of Sokolovski and Baskin to obtain transmission and reflection time distributions with real moments. They make use of a procedure of averaging over times described by Olkhovski and Recami [274] for obtaining the mean time a particle is in a given position in space ('time of presence') and the mean time taken by a particle to traverse a surface ('time of passage'). They also define a 'stay time' as the distribution of the time spent by a particle in a given region. They find that the mean stay time and its standard deviation coincide with the real and imaginary parts respectively of the complex time obtained by Sokolovski and Baskin.

Two problems remain to be properly understood within this technique: the weak dependence of the transmission time on the barrier width and hence the possibility of superluminal traversal speeds, and non-zero and non-positive definite reflection times for regions on the far side of a barrier. The first one could be due to the use of non-relativistic quantum mechanics; it could also be due to quantum nonlocality. A relativistic formulation should be able to resolve the issue, but that is not straightforward. It is possible that the second one will be resolved by using smooth-edged perturbative potentials.

This method has also been extended to three-dimensional systems and multi-channel scattering.

10.3.3   *Wigner trajectories*

It is possible to define trajectories in phase space, called Wigner trajectories [275], based on the Weyl–Wigner representation of quantum mechanics. The Weyl transform of an operator $\hat{A}$ is defined by

$$A_W(x, p) \equiv \int_{-\infty}^{+\infty} \left\langle x - \frac{y}{2} | \hat{A} | x + \frac{y}{2} \right\rangle e^{ipy/\hbar} \, dy. \tag{10.48}$$

The Wigner function f^q is just $1/2\pi\hbar$ times the Weyl transform of the density operator $\hat{\rho}$:

$$f^q(x, p, t) = \frac{1}{2\pi\hbar} \rho_W(x, p, t). \tag{10.49}$$

The trace of the product of two operators $\hat{A}$ and $\hat{B}$ is given by

$$\text{Tr}(\hat{A}, \hat{B}) = h^{-1} \int\int dp \, dx \, A_W(x, p) B_W(x, p), \tag{10.50}$$

where the integrals are from $-\infty$ to $+\infty$. Therefore, the expectation value of $\hat{A}$ is given by

$$\langle \hat{A} \rangle = \int\int f^q(x, p, t) \, A_W(x, p) \, dp \, dx. \tag{10.51}$$

This is of the same form as in classical statistical mechanics, with A_W and f^q playing the roles of the classical function $A(x, p)$ and the classical distribution function f^{cl} respectively. One can therefore build a classical ensemble at time $t = 0$ with $f^{cl}(t = 0) = f^q(t = 0)$. This method has been extensively used in quantum chemistry, quantum optics and quantum chaos for computations. However, although f^q is real by definition, it can be negative in certain regions of phase space. This is why, unlike f^{cl}, it cannot be interpreted as a probability distribution.

The time evolution of the classical distribution f^{cl} is governed by the Liouville equation

$$f_t^{cl} = -f_x^{cl} \frac{p}{m} + f_p^{cl} V_x, \tag{10.52}$$

where $f_X \equiv \partial f/\partial X$, $X = t, p, x$. Differentiating (10.49), one obtains

$$f_t^q = -f_x^q \frac{p}{m} + \frac{2}{\hbar} \sin\left(\frac{\hbar}{2} \frac{\partial}{\partial p} \frac{\partial}{\partial x}\right) V f^q, \tag{10.53}$$

where $\partial/\partial x$ acts only on V. One can therefore formally write

$$f_t^q = -f_x^q \frac{p}{m} + \tilde{V}_x f_p^q \tag{10.54}$$

with

$$\tilde{V}_x \equiv \frac{f_t^q + f_x^q (p/m)}{f_p^q}. \tag{10.55}$$

In formal analogy with classical phase space trajectories, Lee and Scully [275] defined Wigner trajectories through the modified Hamiltonian equations

$$\frac{\mathrm{d}x^{\mathrm{W}}}{\mathrm{d}t} = \frac{p^{\mathrm{W}}}{m},$$
$$\frac{\mathrm{d}p^{\mathrm{W}}}{\mathrm{d}t} = -\frac{\partial \tilde{V}(x, p, t)}{\partial x}\big|_{\{x^{\mathrm{W}}, p^{\mathrm{W}}, t\}}, \tag{10.56}$$

which provide a pictorial representation of the quantum dynamics, $-\tilde{V}_x$ playing the role of a 'quantum force'.

However, Sala, Brouard and Muga [276] have shown in the wave-packet case that Wigner trajectories satisfy Liouville's theorem only locally, i.e., for restricted domains of phase space and time – they can be destroyed or created at singularities of the 'quantum force'. This limits their possible applications to computing quantum averages, particularly tunneling times. They concluded, however, that Weyl transforms of Heisenberg operators are viable alternatives to Wigner trajectories.

10.3.4 Wave-packet trajectories

Although wave packets may be localized, there are inherent ambiguities in defining tunneling times for them, depending on whether one considers the motion of the peak, the centroid, the leading edge, etc. These questions also arise for classical wave propagation, and could equally well be asked in the context of water waves, sound waves, radio waves, and so on. If one tracks the evolution of their peaks or centroids, one often encounters superluminal speeds. This is because an incoming peak or centroid does not in any obvious causal sense evolve into an outgoing peak or centroid. Take, for example, the case of a gaussian minimal uncertainty packet, prepared a long distance from a barrier and travelling towards it. Since the propagation is dispersive, the higher momentum components will not only reach the barrier earlier, they will also be transmitted more effectively than the others because of their higher energies. So, the relevant parameters could be so fixed that the peak or centroid of the transmitted packet emerges from the barrier *before* the incident peak hits the barrier. This shows that there is no causal relationship between these two events [263]. This 'pulse reshaping' due to the preferential attenuation of the later parts of an incident pulse by the medium is known to occur in classical optics [277] .

However, for an alternative viewpoint regarding wave packet trajectories, see Barker *et al.* [278].

10.4   Quantum clocks

An important aspect of the study of tunneling times has been the coupling
of the translational motion of the tunneling particle with an additional
degree of freedom that can act as a clock. Two types of clocks have mainly
been proposed: the Larmor clock [279] and the oscillating barrier [255].

The most widely studied is the Larmor clock. If a weak magnetic
field is applied and confined to the barrier region, then the state of an
incident particle with a spin S will be a linear superposition of $(2S + 1)$ spin states, each with a different energy due to the Zeeman effect.
These states will therefore have different transmission rates, leading to
an effective precession and rotation of the spin whose components can
then act as measures of the traversal times. As already mentioned, these
times correspond in the BSM approach to different combinations of
the projectors P and D, and in the Feynman path-integral approach of
Sokolovski and Baskin to the real and imaginary parts of complex times.
Let the incident particle have spin $\frac{1}{2}$ and move along the x-direction with
its spin polarized in the y-direction, and let the uniform magnetic field be
applied along the z-direction. Then the amount of spin precession in the
x–y plane and 'rotation' into the field direction clock the characteristic
tunneling times τ_x and τ_z respectively. The Büttiker–Landauer time is
$\tau = \sqrt{\tau_x^2 + \tau_y^2}$.

Much of the interest in tunneling times was, in fact, stimulated by
the papers of Landauer and Büttiker who studied tunneling through a
rectangular barrier with a small oscillating component added to the height,

$$V(t) = V_0 + V_1 \cos \omega t. \tag{10.57}$$

The incident particles with energy E can absorb or emit modulation
quanta $\hbar\omega$ during tunneling, leading to the appearance of sidebands with
energies $E \pm \hbar\omega$ to first order in V_1. For an opaque barrier with frequencies
$\hbar\omega$ small compared to E and $V_0 - E$, the relative sideband intensities are

$$I_\pm^T(\omega) \equiv |A_\pm(\omega)/A_0|$$
$$= \left[\frac{V_1}{2\hbar\omega}\right]^2 \left[\exp\left(\pm\omega\frac{md}{\hbar\kappa}\right) - 1\right]^2, \tag{10.58}$$

where $A_\pm$ are the sideband transmission amplitudes, and A_0 is the transmission amplitude in the absence of the perturbation. It is clear that $\hbar\kappa/md$
is the characteristic frequency separating the two sidebands. Büttiker and
Landauer identified $md/\hbar\kappa$ with the traversal time for tunneling. Hauge
and Stovneng [259] have, however, shown by the example of the δ-function
barrier that no direct relation exists between the traversal time and the
characteristic frequency of an oscillating barrier. For barriers of inter-

mediate thickness a characteristic frequency cannot even be meaningfully defined. The proper meaning of the Büttiker–Landauer time τ_T is obtained from the work of Iannaccone and Pellegrini [273] – it is the root mean square transmission time. This accounts for τ_T and τ_R each being *greater* than the dwell time τ_D, a feature inexplicable in the original interpretation.

10.5 Experiments

The experimental situation has been reviewed by Landauer [280] and Landauer and Martin [263]. The remarkable fact is that a time scale associated with barrier traversal can be measured and is real, whatever its relationship with theoretical controversies might be. We will discuss only the principles of three different kinds of experiments that have been done so far.

In the first type, low-barrier heterostructures have been systematically investigated over a wide range of barrier heights and widths. While tunneling, the electron is attracted back by an electrostatic image charge on the electrode it is leaving. The extent to which this image charge has time to spread can be used as a measure of the tunneling time. Guéret *et al.* [281] found experimental evidence for the validity of the static image correction for large tunneling times, and for the occurrence of dynamic effects in the limit of short tunneling times. However, the data were not accurate enough to discriminate between the various alternative theoretical predictions.

The second type of experiment [282] (Fig. 10.1) used a Josephson junction passing a constant current J less than its critical current J_0, i.e., the maximum current it can support without a voltage. The current is supplied from a source with a capacitance C. The junction is connected to a terminating resistor Z_t through a transmission line of impedance Z_l and of variable length l and therefore of variable delay. The energy U as a function of θ, the phase difference across the junction, is given by

$$U(\theta) = (J_0 \cos \theta - J\theta)\, \hbar/2e. \qquad (10.59)$$

This potential has local minima if $|J| < J_0$, and the circuit can tunnel out of these local metastable states. (Fig. 10.2). For $J > 0$ resistive dissipation lowers the energy of the system, and tunneling occurs accompanied by the generation of a voltage $V = (\hbar/2e)\, d\theta/dt$ which can be measured. After each decay event, the circuit can be reset, and the time elapsed until the next decay measured.

The resistor Z_t is chosen to be much smaller than Z_l, so that the voltage pulses associated with the tunneling process travel towards it and are reflected back to the Josephson junction. If the length l is so adjusted that the time taken by the voltage pulse to reach Z_t is longer

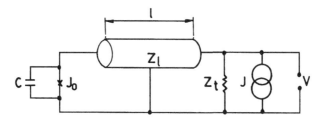

Fig. 10.1. Circuitry of the experiment by Esteve *et al.* (after Ref. [280]).

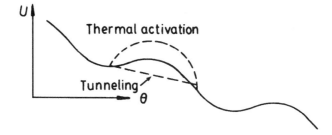

Fig. 10.2. Metastable states of a circuit containing a Josephson junction (after Ref. [280]).

than the tunneling time, the energy losses (which reduce the tunneling probability [283]) are determined by Z_l alone. On the other hand, if the tunneling time is so long that the voltage pulses are able to reach Z_t and get reflected back to the junction while the tunneling occurs, the energy losses are determined by Z_t as well. Esteve *et al.* studied the cross-over between the two situations and claimed to have found a time consistent with the Büttiker–Landauer time. This interpretation, however, depends on the formal analogy that the Josephson junction 'can be modelled as a particle of coordinate δ [which is in reality, though, the phase difference between the layers of the junction] moving in a one-dimensional tilted washboard potential' [282]. As argued by Cushing [267], it is precisely because this variable δ actually represents a collective phenomenon for an entire circuit that the scale for its change is long enough to make it measurable. It is therefore of doubtful relevance for the times defined for the scattering configuration where a particle is incident on a barrier and tunnels through it. It is perhaps more relevant for tunneling times of metastable bound states which we have not discussed.

Nevertheless, both these experiments provide examples of clocks sensitive to the system's interaction time with the barrier – surface charges in the case of electrons and transmission line modes in the case of a Josephson junction, which may or may not have time to adjust to the

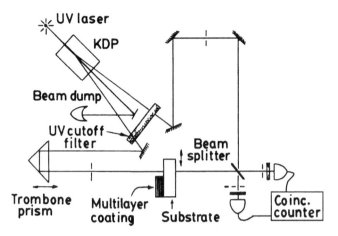

Fig. 10.3. Apparatus for measuring the single-photon tunneling time (after Ref. [284]).

system's motion through the barrier. The measured times in the first case are of the order of 10^{-14} or 10^{-15} seconds, typical of atomic distances and electron velocities, whereas in the second case they are about 80 picoseconds which is much longer and can therefore be measured with the help of a mechanically variable transmission line.

The third experiment involves measurement of single-photon tunneling times [284]. Although an understanding of electron tunneling is currently more important for actual devices, single-photon tunneling is more convenient for unravelling the fundamental problems in our understanding of the tunneling process. There are two reasons for this. First, the wavelength of light being much larger than that of electrons in solids, the size of the barrier can be typically much larger for photon tunneling than for electron tunneling. Second, it is possible to use quantum optical techniques to perform high-resolution measurements of the arrival times of individual photons.

A pair of parametrically down-converted photons is produced and made to pass through a modified Hong–Ou–Mandel interferometer [285] (Fig. 10.3). One of the photons passes through air and the other through a sample consisting of a substrate of fused silica, half of whose face is coated with a 1.1 μm thick multi-layer dielectric mirror made from quarter-wave layers of alternating high- and low-index materials, which acts like a barrier for a range of frequencies. The entire opposite face of the sample is anti-reflection coated. The sample is mounted on a precision translation stage, and can be placed in either of two positions: in one of these positions the photon must tunnel through the 1.1 μm coating

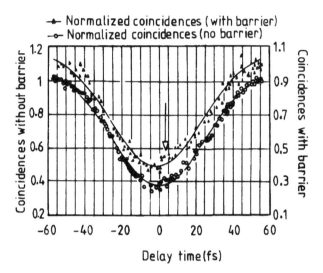

Fig. 10.4. Coincidence profiles with and without the tunnel barrier, taken by scanning the trombone prism, map out the single-photon wave packets. The upper curve (right axis) shows the coincidences with the barrier; this curve is shifted by 1.1 ± 0.3 fs to *negative* times relative to the one with no barrier (lower curve, left axis). (For comparison, the arrow corresponds to the *delay* time one would expect from the optical path length of the coating divided by c, the velocity of light (after Ref. [284]).

in order to be transmitted, while in the other position it travels through 1.1 μm of air. The two conjugate photons are brought back together by means of mirrors so that they arrive simultaneously on the surface of a 50/50 beam splitter. Coincidence counts are recorded when detectors placed at the two output ports of the beam splitter register counts within 500 picoseconds of one another.

If the wave packets of the two photons overlap in time at the beam splitter, a destructive interference effect leads to a null in the coincidence detection rate [286]. As the path-length difference is changed by translating a 'trombone' prism, the coincidence rate exhibits a dip (with an r.m.s. width of approximately 20 fs, the correlation time of the two photons) corresponding to the perfect overlap of the two wave packets. When the barrier (the 1.1 μm coating) is inserted into one arm, a path difference is introduced, which can be compensated by translating the trombone prism. It was found that the external delay had to be *lengthened*, implying that the barrier had speeded up the photon (Fig. 10.4). Thirteen fits yielded a relative delay of -1.47 ± 0.21 fs, including the estimated systematic errors. It is clear from Fig. 10.2 that this is yet another example of the 'pulse reshaping' phenomenon mentioned earlier in connection with wave-packet trajectories. Similar results were also obtained earlier using microwaves by Ranfagni *et al.* [287] and Enders and Nimtz [288], but their measurements suffered from technical complexities related to the exact positions of the microwave probes and effects of geometrical discontinuities.

A clear advantage of this type of experiment with simultaneously generated and EPR correlated photon pairs over electron tunneling experiments is that after one photon traverses a tunnel barrier, its time of arrival at the detector can be compared with that of its twin which encounters no barrier, offering a clear operational definition of the tunneling time. Moreover, the magnitude of this time difference is so small as to be inaccessible to electronic measurements but can be measured by the interference technique.

The experiment showed that single photons travel at the group velocity in glass! Moreover, since the technique relies on coincidence detection, the *particle aspect* of tunneling (essentially a *wave phenomenon*) can be clearly observed: each coincidence detection corresponds to a *single* tunneling event.

Unfortunately, these interpretations depend on a *formal* analogy between the Helmholtz equation, the equation for steady-state electromagnetic waves in an inhomogeneous but isotropic medium in the scalar approximation, and the time-independent Schrödinger equation for massive particles [286]. Any photon–particle analogy made to relate optical delay times to particle tunnelling time is suspect in the absence of a consistent particle interpretation of the photon. Such an interpretation has recently been proposed [43], and it would be interesting to calculate single-photon tunnelling times using this formalism and compare the results with experiments.

References

[1] Pauli, W. (1959). *Einstein's Contributions to Quantum Theory* in *Albert Einstein: Philosopher-Scientist*, ed. P. A. Schilpp (Library of Living Philosophers Inc., Harper & Row, New York, Evanston and London, 1959 Harper Torchbooks Science Library edition), p. 85.

[2] Ghose, P. (1994). *Bose Statistics: A Historical Perspective* in *S. N. Bose: The Man And His Work, Part I*, ed. C. K. Majumdar et al. (S. N. Bose National Centre for Basic Sciences, Calcutta), p. 35.
Pais, A. (1982). *Subtle is the Lord...* (Oxford University Press, Oxford).
Jammer, M. (1966). *Conceptual Development of Quantum Mechanics* (McGraw-Hill Book Co.).

[3] Combourie, Marie-Christine & Rauch, H. (1992). *Found. of Phys.* **22**, 1403.

[4] Bohr, N. (1949). *Discussion with Einstein on Epistemological Problems in Atomic Physics* in Schilpp [1], p. 199.

[5] Scully, M. O., Englert, B. & Walther, H. (1991). *Nature* **351**, 111.

[6] Storey, P. *et al.* (1994). *Nature* **367**, 626.

[7] Mandel, L. (1991). *Opt. Lett.* **16**, 1882.

[8] Mandel, L. (1995). Rochester preprint.

[9] Englert, B.-G. (1996). *Phys. Rev. Lett.* **77**, 2154.

[10] Horne, M. A., Shimony, A. & Zeilinger, A. (1989). *Phys. Rev. Lett.* **19**, 2209.
Jaeger, G., Horne, M. A. & Shimony, A. (1993). *Phys. Rev.* **A48**, 1023.

[11] Zou, X. Y., Wang, L. J. & Mandel, L. (1991). *Phys. Rev. Lett.* **67**, 318.

[12] Ray, A. & Home, D. (1995). *Phys. Lett.* **A204**, 87.

[13] Rauch, H., Trimer, W. & Bonse, U. (1974). *Phys. Lett.* **A47**, 369.

[14] Rauch, H. & Summhammer, J. (1984). *Phys. Lett.* **A104**, 44.

[15] Wooters, W. K. & Zurek, W. H. (1979). *Phys. Rev.* **D19**, 473.

[16] Greenberger, D. M. & Yasin, A. (1988). *Phys. Lett.* **A128**, 391.

[17] Home, D. & Kaloyerou, P. N. (1989). *J.Phys.* **A22**, 3253.

[18] Badurek, G., Rauch, H. & Tuppinger, D. (1986). *Phys. Rev.* **A34**, 2600.

[19] Kaiser, H., Werner, S. A. & George, E. A. (1983). *Phys. Rev. Lett.* **50**, 560.

[20] Werner, S. A., Clothier, R., Kaiser, H., Rauch, H. & Wölwitsch, H. (1991). *Phys. Rev. Lett.* **67**, 683.
Kaiser, H., Clothier, R., Werner, S. A., Rauch, H. & Wölwitsch, H. (1992). *Phys. Rev.* **A45**, 31.
Rauch, H. (1993). *Phys. Lett.* **A173**, 240.
Jacobson, D. L., Werner, S. A. & Rauch, H. (1994). *Phys. Rev.* **A49**, 3196.

[21] Ray, A. & Home, D. (1993). *Phys. Lett.* **A178**, 33.

[22] Tonomura, A. (1990). *Foundations of Quantum Mechanics in the Light of New Technology*, Proc. of the 3rd International Symposium (Physical Society of Japan, Tokyo), p. 15.
Tonomura, A., Endo, J., Matsuda, T., Kawasaki, T. & Ezawa, H. (1989). *Am. J. of Phys.* **57**, 117.

[23] Holland, P. R. (1993). *The Quantum Theory Of Motion* (Cambridge University Press, Cambridge) and references therein.

[24] Selleri, F. (1990). In *Quantum Paradoxes and Physical Reality*, ed. A. Van der Merwe (Kluwer Academic, Dordrecht); (1992). *The Wave-Particle duality* (Plenum, New York).

[25] Wang, L. J., Zou, X. Y. & Mandel, L. (1991). *Phys. Rev. Lett.* **66**, 1111.

[26] Holland, P. R. & Vigier, J. P. (1991). *Phys. Rev. Lett.* **67**, 402.

[27] Grangier, P., Roger, G. & Aspect, A. (1986). *Europhys. Lett.* **1**,173.

[28] Aspect, A. & Grangier, P. (1987). *Hyperfine Int.* **37**, 3.

[29] Aspect, A. (1990). In *Sixty-two Years of Uncertainty*, ed. A. I. Miller (Plenum, New York), p. 45.

[30] Taylor, G. I. (1909). *Proc. Camb. Phil. Soc.* **15**, 114.
See also Janossy, L. & Narray, Z. (1967). *Acta Phys. Hungarica* **7**, 403 and Ref. 20.

[31] Clark, R. W. 91971). In *Einstein : The Life And Times* (World Publishing Co., New York); (Avon Books, paparback edition, New York, 1972).

[32] Loudon, R. (1983). *The Quantum Theory of Light* (Clarendon, Oxford).

[33] Fry, E. S. (1973). *Phys. Rev.* **A8**, 1219.

[34] Bohr, N. (1949). (See ref. [4], p. 203).

[35] Ghose, P., Home, D. & Agarwal, G. S. (1991). *Phys. Lett.* **A153**, 403.; (1992). *ibid* **A168**, 95.

[36] Sommerfeld, A. (1964). *Optics* (Academic Press, New York), pp. 32–3.

[37] Mizobuchi, Y. & Ohtaké, Y. (1992). *Phys. Lett.* **A168**, 1.

[38] Burnham, D. C. & Weinberg, D. L. (1970). *Phys. Rev. Lett.* **25**, 84.

[39] Unnikrishnan, C. S. & Murthy, S. (1996). *Phys. Lett.* **A221**, 1.

[40] Fizeau, A. H. (1890). *Comptes Rendus* **29**, 90.

[41] Einstein, A. (1905). *Ann. Phys. (Leipzig)* **18**, 639.
de Broglie, L. (1926). *C. R. Acad. Sci. (Paris)* **183**, 447; (1927). *C. R. Acad. Sci. (Paris)* **185,** 580.
Bohm, D. (1952). *Phys. Rev.* **85**, 166, 180.

[42] Dewdney, C., Horton, G., Lam, M. M., Malik, Z. & Schmidt, M. (1992). *Found. of Phys.* **22**, 1217.

[43] Ghose, P. (1996). *Found. of Physics* **26**, 1441, and references therein.

[44] Prosser, R. D. (1976). *Int. J. of Theoret. Phys.* **15**, 169, 181.

[45] Einstein, A. (1916). *Verh. d. Deutsch. Phys. Ges. (2)* **18**, 318; *Mitteilungen der Phys. Ges. Zürich* **16**, 47.

[46] Einstein, A. (1917). *Phys. Zs.* **18**, 121.

[47] Haroche, S. (1992). *Cavity Quantum Electrodynamics* in *Fundamental Systems in Quantum Optics*, Les Houches, Session LIII, 1990, eds. J. Dalibard, J. M. Raimond & J. Zinn-Justin (Elsevier Science Publishers, New York). Other references to the literature will be found here.

[48] Cohen-Tannoudji, C. (1984). In *New Trends in Atomic Physics*, Les Houches Summer School Session *XXXVIII*, eds. G. Grynberg & R. Stora (North-Holland, Amsterdam).

[49] Fain, V. M. (1966). *Sov. Phys. JETP* **23**, 882.
Fain, V. M. & Khanin, Y. I. (1969). *Quantum Electronics* (MIT Press, Cambridge M. A.).
Fain, V. M. (1982). *Il Nuovo Cimento* , **68B**, 73.

[50] Dalibard, J. Dupont-Roc, J. & Cohen-Tannoudji, C. (1982). *J. Phys. (Paris)* **43**, 1617.

[51] Meschede, D., Jhe, W. & Hinds, E. A. (1990). *Phys. Rev.* **A41**, 1587.
Jhe, W. (1991). *ibid* **A43**, 4199.

[52] Drexhage, K. H. (1974). In *Progress in Optics XII*, ed. E. Wolf (North-Holland, Amsterdam).

[53] Kleppner, D. (1981). *Phys Rev. Lett.* **47**, 233.

[54] Brune, M., Schmidt-Kaler, F., Maali, A., Dreyer, J., Hagley, E., Raimond, J. M. & Haroche, S. (1996). *Phys. Rev. Lett.* **76**, 1800.

[55] Kimble, H. J., Dagenais, M. & Mandel, L. (1977). *Phys. Rev. Lett.* **39**, 691.

[56] Cresser, J. D., Hägger, J., Leuchs, G., Rateike, M. & Walther, H. (1982). *Dissipative Systems in Quantum Optics* **21** (Springer, Berlin).

[57] Short, R. & Mandel, L. (1983). *Phys. Rev. Lett.* **51**, 384.

[58] Slusher, R. E., Hollberg, L. W., Yurke, B., Mertz, J. C. & Valley, J. F. (1985). *Phys. Rev. Lett.* **55**, 2409.

[59] Loudon, R. & Knight, P. L. (1987). *J. Mod. Opt.* **34**, 707.

[60] Brune, M., Haroche, S., Lefevre, V., Raimond, J. M. & Zagury, N. (1990). *Phys. Rev. Lett.* **65**, 976.

[61] Brune, M., Nussenzveig, P., Schmidt-Kaler, F., Bernardot, F., Maali, A., Raimond, J. M. & Haroche, S. (1994). *Phys. Rev. Lett.* **72**, 3339.

[62] Barenco, A. Deutsch, D., Ekert, A. & Josja, R. (1995). *Phys. Rev. Lett.* **74**, 4083.
Sleator, T. & Weinfurter, H. (1995). *Phys. Rev. Lett.* **74**, 4087.
Domokos, P., Brune, M., Raimond, J. M. & Haroche, S. (1995). *Phys. Rev.* **A52**, 3554.

[63] Schrödinger, E. (1935).*The Present Situation in Quantum Mechanics* (English translation), *Proc. of the American Phys. Soc.*, **124**, 323.
Reprinted in *Quantum Theory and Measurement*, eds. J. A. Wheeler & W. H. Zurek (Princeton University Press, Princeton, New Jersey, 1983).

[64] Glauber, R. J. (1986) in *Frontiers in Quantum Optics*, ed. E. R. Pike (Hilger, London), p. 534.

[65] Yurke, B. & Stoler, D. (1986). *Phys. Rev. Lett.* **57**, 13.
Milburn, G. J. & Holmes, C. A. (1986). *ibid* **56**, 2237.
Bužek, V., Moya-Cessa, H., Knight, P. L. & Phoenix, S. J. D. (1992). *Phys. Rev.* **A45**, 8190.
Mecozzi, A. & Tombesi, P. (1987). *Phys. Rev. Lett.* **58**, 1055.
Tombesi, P, & Mecozzi, A. (1987). *J. Opt. Soc. Am.* **B4**, 1700.

[66] Tara, K., Agarwal, G. S. & Chaturvedi, S. (1993). *Phys. Rev.* **A47**, 5024.
Gantsog, Ts. & Tanas, R. (1991). *Quantum Opt.* **3**, 33.

[67] Slosser, J. J., Meystre, P. & Wright, E. M. (1990). *Opt. Lett.* **15**, 233.
Slosser, J. J. & Meystre, P. (1990). *Phys. Rev.* **A41**, 3867.
Wilkens, M. & Meystre, P. (1991). *ibid* **A43**, 3832.
Meystre, P., Slosser, J. J. & Wilkens, M. (1991). *ibid* **A43**, 4959.

[68] Song, S., Caves, C. M. & Yurke, B. (1990). *Phys. Rev.* **A41**, 5261.
Yurke, B., Schleich, W. & Walls, D. F. (1990). *ibid* **A42**, 1703.
Brune, M., Haroche, S., Raimond, J. M., Davidovich, L. & Zagury, N. (1992). *ibid* **A45**, 5193.

[69] Agarwal, G. S. & Tara, K. (1992). *Phys. Rev.* **A46**, 485.
Varada, G. V. & Agarwal, G. S. (1993). *ibid* **A48**, 4062.

[70] Brune, M., Hagley, E., Dreyer, J., Maître, X., Maali, A., Wunderlich, C., Raimond, J. M. & Haroche, S. (1996). *Phys. Rev. Lett.* **77**, 4887.

[71] Monroe, C., Meekhof, D. M., King, B. E. & Wineland, D. J. (1996). *Science* **272**, 1131.

[72] Scully, M. O. & Drühl, K. (1982). *Phys. Rev.* **A25**, 2208.

[73] Herzog, T. J., Kwiat, P. G., Weinfurter, H. & Zeilinger, A. (1995). *Phys. Rev. Lett.* **75**, 3034.

[74] Braginsky, V. B., Vorontsov, Yu. I. & Khalili, F. Ya, (1977). *Sov. Phys. JETP* **46**, 705.

[75] Unruh, W. G. (1978). *Phys. Rev.* **D18**, 1764.

[76] Caves, C. M., Thorne, K. S., Drever, R. W. P., Sandberg, V. D. & Zimmermann, M. (1980). *Rev. Mod. Phys.* **52**, 341.

[77] Braginsky, V. B., Vorontsov, Yu. I. & Thorne, K. S. (1980). *Science* **209**, 547. (Reprinted in Wheeler & Zurek, 1983).

[78] Braginsky, V. B. & Vorontsov, Yu. I. (1974). *Usp. Fiz. Nauk.* **114**, 41. [(1975.*Sov. Phys. Usp.* **17**, 644.]

[79] Gol'dman, I. I. & Krivchenkov,V. D. (1961). *Problems in Quantum Mechanics* (Pergamon, London).

[80] Hollenhorst, J. N. (1979). *Phys. Rev.* **D19**, 1669.

[81] Milburn, G. J. & Walls, D. F. (1983). *Phys. Rev.* **A28**, 2065.

[82] Imoto, N., Haus, H. A. & Yamamoto, Y. (1985). *Phys. Rev.* **A32**, 2287.

[83] Levenson, M. D., Shelby, R. M., Reid, M. & Walls, D. F. (1986). *Phys. Rev. Lett.* **57**, 2473.

[84] La Porta, A., Slusher, R. E. & Yurke, B. (1989). *Phys. Rev. Lett.* **62**, 28.

[85] Grangier, P., Roch, J. & Roger, G. (1991). *Phys. Rev. Lett.* **66**, 1418.

[86] Aharonov, Y. & Bohm, D. (1959). *Phys. Rev.* **115**, 485; (1961). *ibid* **123**, 1511; (1962);*ibid* **125**, 2192; (1963). *ibid* **130**, 1625.

[87] Ehrenberg, W. & Siday, R. E. (1949). *Proc. Phys. Soc., London* **B62**, 8.

[88] Bocchieri, P., & Loinger, A. (1978). *Nuovo Cimento* **47A**, 475; (1979). *Lett. Nuovo Cimento* **25**, 476; (1980); *Nuovo Cimento* **59A**, 121; (1981). *ibid* **66A**, 144.
Bocchieri, P., Loinger, A. & Siragusa, G. (1980). *ibid* **56A**, 55.
Bocchieri, P. & Loinger, A. (1981). *ibid* **66A**, 164.

[89] Home, D. & Sengupta, S. (1983). *Am. J. Phys.* **51**, 942.

[90] Furry, W. H. & Ramsey, N. (1960). *Phys. Rev.* **118**, 623.

[91] Wu, T. T. & Yang, C. N. (1975). *Phys. Rev.* **D12**, 3843, 3845.

[92] Strocchi, F. & Wightman, A. S. (1974). *J. Math. Phys.* **15**, 2198.
Casati G. & Guarneri, I. (1979). *Phys. Rev. Lett.* **42**, 1578.

[93] Werner, F. G. & Brill, D. R. (1960). *Phys. Rev. Lett.* **4**, 349.
Chambers, R. G. (1960). *ibid* **5**, 3.
Fowler, H. A., Marton, L., Simpson, Y. A. & Suddeth, J. A. (1961). *J. Appl. Phys.* **32**, 1153.
Möllenstedt, G. & Bayh, W. (1961). *Naturwiss.* **48**, 400; (1962). *ibid* **49**, 81.
Bayh, W. (1962). *Z. Phys.* **169**, 492.

Boersch, H., Hamisch, H., Wohlleben, D. & Grohmann, K. (1960). *ibid* **159**, 397; (1961). *ibid*, **165**, 79.
Boersch, H., Hamisch, H. & Grohmann, K. (1962). *ibid*, **169**, 263.
Jaklevic, R. C., Lambe, J., Mercereau, J. E. & Silver, A. H. (1964). *Phys. Rev. Lett.* **12**, 274; (1965). *Phys. Rev.* **A140**, 1628.
Lischke, B. (1969). *Phys. Rev. Lett.* **22**, 1366.
Matteucci, G., Missiroli, G. F. & Pozzi, G. (1977). *G. Fis.* **18**, 264.

[94]   Roy, S. M. (1980). *Phys. Rev. Lett.* **44**, 111.

[95]   Tonomura, A., Matsuda, T., Suzuki, R., Fukuhara, A., Osakabe, N., Umezaki, H., Endo, J., Shinagawa, K., Sugita, Y. & Fujiwara, H. (1982). *Phys. Rev. Lett.* **48**, 1443.

[96]   Costa de Beauregard, O. Vigoureux, J. M. (1974). *Phys. Rev.* **D9**, 1626; (1982). *Lett. al Nuovo Cimento* **33**, 79.
Kulik, O. (1970). *JETP Lett.* **11**, 275.

[97]   Tonomura, A., Umezaki, H., Matsuda, T., Osakabe, N., Endo, J. & Sugita, Y. (1983). *Phys. Rev. Lett.* **51**, 331.

[98]   Allman, B. E., Cimmino, A., Klein, A. G. & Opat, G. I. (1992). *Phys. Rev. Lett.* **68**, 2409.

[99]   Anandan, J. S. (1982). *Phys. Rev. Lett.* **48**, 1660.

[100]   Aharonov, Y. & Casher, A. (1984). *Phys. Rev. Lett.* **53**, 319.

[101]   Anandan, J. (1988). *Phys. Lett.* **A133**, 171;
(1989). *ibid* **A138**, 347;
(1994). *ibid* **A141**, 335;
(1990). *Proc. of the Int. Symposium on the Foundations of Quantum Mechanics*, Tokyo, August 1989, eds. S. Kobayashi *et al* (Physical Society of Japan), pp. 98–106;
(1994). *Topological Phases and Duality in Electromagnetic and Gravitational Fields* in *Quantum Coherence and Reality*, Proc. of the Conference on *Fundamental Aspects of Quantum Theory*, Columbia, SC, Dec. 1992, eds. J. S. Anandan & J. L. Safko (World Scientific, Singapore).

[102]   Ramsey, N. F. (1993). *Phys. Rev.* **A48**, 80.

[103]   Goldhaber, A. S. (1989). *Phys. Rev. Lett.* **62**, 482.

[104]   Boyer, T. H. (1987). *Phys. Rev.* **A36**, 5083.

[105]   Cimmino, A., Opat, G. I., Klein, A. G., Kaiser, H., Werner, S. A., Arif, M. & Clothier, R. (1989). *Phys. Rev. Lett.* **63**, 380.

[106]   Aharonov, Y. & Anandan, J. (1987). *Phys. Rev. Lett.* **58**, 1593.

[107]   Simon, B. (1983). *Phys. Rev. Lett.* **51**, 2167.

[108]   Berry, M. V. (1984). *Proc. Roy. Soc. London* **A392**, 45.
Pancharatnam, S. (1956). *Proc. Indian Acad. Sci.* **A44**, 247.

[109]   Anandan, A. & Aharonov, Y. (1988). *Phys. Rev.* **D38**, 1863.

[110] Samuel, J. & Bhandari, R. (1988). *Phys. Rev. Lett.* **60**, 2339.
Mukunda, N. & Simon, R. (1993). *Ann. of Phys.* **228**, 205, 269.

[111] *Geometric Phase in Physics*, eds. A. Shapere & F. Wilczek (World Scientific, Singapore).

[112] von Neumann, J. (1933). *Mathematical Foundations of Quantum Mechanics*, translated by R. T. Beyer (Princeton University Press, 1955), chapters 5 and 6.

[113] Everett, H.III (1957). *Reviews of Modern Physics*, **29**, 454. (Reprinted in Wheeler & Zurek, 1983).

[114] Daneri, A., Loinger, A. & Prosperi, G. M. (1962). *Nucl. Phys.* **33**, 297.
Bub, J. (1968). *Nuovo Cimento* **57B**, 503.
Moldauer, P. A. (1972). *Phys. Rev.* **D5**, 1028.

[115] Joos, E. & Zeh, H. D. (1985). *Z. Phys.* **B59**, 223.

[116] Zurek, W. H. (1981). *Phys. Rev.* **D24**, 1516.

[117] Zurek, W. H. (1982). *Phys. Rev.* **D26**, 1862.

[118] Zurek, W. H. (1991). *Phys. Today* October, p. 36.

[119] Anderson, J. L. (1993). *Phys. Today* April, p. 13.

[120] Ghirardi, G. C., Grassi, R. & Pearle, P. (1993). *Phys. Today* April, p. 13.

[121] Hepp, K. (1993). *Phys. Today* April, p. 15.

[122] Gell-Mann, M. & Hartle, J. B. (1991). *Proc. of the 25th Int. Conf. on HEP*, eds. K. K. Phua & Y. Yamaguchi (South East Asia Theoretical Physics Association & The Physical Society, Japan) and references therein

[123] Gell-Mann, M. & Hartle, J. B. (1993). *Phys. Rev.* **D47**, 3345.

[124] Gell-Mann, M. & Hartle, J. B. (1990). In *Complexity, Entropy, and the Physics of Information*, Santa Fe Institute Studies in the Sciences of Complexity VIII, ed. W. H. Zurek (Addison-Wesley, Reading) and in *Proc. of the 3rd International Symposium on the Foundations of Quantum Mechanics in the Light of New Technology*, eds. S. Kobayashi, H. Ezawa, Y. Murayama & S. Nomura (The Physical Society of Japan, Tokyo).

[125] Griffiths, R. (1984). *J. Stat. Phys.* **36**, 219.

[126] Omnès, R. (1988). *J. Stat. Phys.* **53**, 893; *ibid.*, **53**, 933; *ibid.* **53**, 957.

[127] Omnès, R. (1990). *Ann. Phys.* (N.Y.) **201**, 354.

[128] Bohm, D. (1951). *Quantum Theory* (Dover Publications, Inc., New York, 1989), chapter 22, pp. 600–1.

[129] Bohm, D. (1952). *Phys. Rev.* **85**, 166, 180 (Reprinted in Wheeler & Zurek [63], 1983).

[130] Pearle, P. (1976). *Phys. Rev.* **D13**, 857.

[131] Leggett, A. J. (1980). *Supplement of the Prog. Theoret. Phys.* **69**, 80.

[132] Caldeira, A. O. & Leggett, A. J. (1983). *Ann. Phys.* (New York) **149**, 374; *erratum* (1984) *ibid.* **153**, 445.

[133] Leggett, A. J. (1986). In *Directions in Condensed Matter Physics*, eds. G. Grinstein & G. Mazenko (World Scientific, Singapore), pp. 187–248.

[134] de Bruyn Ouboter, R. & Bol, D. (1982). *Physica* **B15**, 15.

[135] Jackel, L. D., Gordon, J. P., Hu, E. L., Howard, R. E., Fetter, L. A., Tennant, D. M., Epworth, R. W. & Kukrijärvi, J. (1981). *Phys. Rev. Lett.* **47**, 697.

[136] Voss, R. F. & Webb, R. A. (1981). *Phys. Rev. Lett.* **47**, 265.

[137] Furry, W. H. (1936). *Phys. Rev.* **49**, 393.

[138] Leggett, A. J. & Garg, A. (1985). *Phys. Rev. Lett.* **54**, 857.

[139] Simonius, M. (1978). *Phys. Rev. Lett.* **40**, 980.

[140] Leggett, A. J., Chakravarty, S., Dorsey, A. T., Fisher, M. P. A., Garg, A. & Zwerger, W. (1987). *Rev. Mod. Phys.* **59**, 1.

[141] Ballentine, L. E. (1987). *Phys. Rev. Lett.* **59**, 1403.

[142] Leggett, A. J. & Garg, A. (1987). *Phys. Rev. Lett.* **59**, 1621.

[143] Peres, A. (1988). *Phys. Rev. Lett.* **61**, 2019.

[144] Tesche, C. D. (1990). *Phys. Rev. Lett.* **64**, 2358.

[145] Hardy, L., Home, D., Squires, E. J. & Whitaker, M. A. B. (1992). *Phys. Rev.* **A45**, 4267.

[146] Khalfin, L. A. (1958). *Sov. Phys. JETP* **6**, 1053.

[147] Winter, R. G. (1961). *Phys. Rev.* **123**, 1503.

[148] Allcock, G. R., (1969). *Ann. Phys. (N.Y.)* **53**, 251.

[149] Yorgrau, W. (1968). In *Problems in Philosophy of Science*, eds. I. Lakatos & A. Musgrave (North-Holland, Amsterdam), p. 191.

[150] Ekstein, H. & Seigert, A. (1971). *Ann. Phys. (N.Y.)* **68**, 509.

[151] Misra, B. & Sudarshan, E. C. G. (1977). *J. Math. Phys.* **18**, 756.

[152] Sudbury, A. (1986). In *Quantum Concepts in Space and Time*, eds. R. Penrose & C. J. Isham (Clarendon Press, Oxford), p. 65.

[153] Greenland, P. T. (1988). *Nature* **335**, 298.

[154] Itano, W. M., Heinzen, D. J., Bollinger, J. J. & Wineland, D. J. (1990). *Phys. Rev.* **A41**, 2295.

[155] Home, D. & Whitaker, M. A. B. (1986). *J. Phys.* **A19**, 1847.

[156] Fleming, G. N. (1973). *Nuovo Cimento* **16A**, 232.

[157] Kraus, K. (1981). *Found. of Phys.* **11**, 547.

[158] Norman, E. B., Gazes, S. B., Crane, S. G. & Bennett, D. A. (1988). *Phys. Rev. Lett.* **60**, 2246.

[159] Cook, R. J. (1988). *Phys. Scr.* **T21**, 49.

[160] Dehmelt, H. (1986). *Proc. Natl. Acad. Sci. U.S.A.* **83**, 2291, 3074.

[161] Peres, A. & Ron, A. (1990). *Phys. Rev.* **A42**, 5720.

[162] Petrovsky, T., Tasaki, S. & Prigogine, I. (1990). *Phys. Lett.* **A151**, 109.

[163] Ballentine, A. (1991). *Phys. Rev.* **A43**, 5165.

[164] Home, D. & Whitaker, M. A. B. (1992). *J. of Phys.* **A25**, 657.

[165] Home, D. & Whitaker, M. A. B. (1993). *Phys. Lett.* **A173**, 327.

[166] Inagaki, S., Namiki, M. & Tajiri, T. (1992). *Phys. Lett.* **A166**, 13.

[167] Agarwal, G. S. & Tewari, S. P. (1994). *Phys. Letts.* **A185**, 139.

[168] Hund, L. (1927). *Z. Phys.* **43**, 805.

[169] Harris, R. A. & Stodolsky, L. (1981). *J. Chem. Phys.* **74**, 2145.
Bhandari, R. (1985). *Pramana—J. Phys.* **25**, 377.

[170] Wolfenstein, L. (1978). *Phys. Rev.* **D17**, 2369.
Mikheyev, S. P. & Smirnov, A. Yu. (1985). *Yad. Fiz.* **42**, 1441.

[171] Pearle, P. (1996). *True Collapse and False Collapse* in *The Proceedings of the 4th Drexel Symposium on Quantum Nonintegrability*, eds. Da Hsuan Feng *et al.* (International Press), p. 51.

[172] Ghirardi, G. C., Rimini, A. & Weber, T. (1986). *Phys. Rev.* **D34**, 470.

[173] Pearle, P. (1989). *Phys. Rev.* **A39**, 2277.
Ghirardi, G. C., Pearle, P. & Rimini, A. (1990). *Phys. Rev.* **A42**, 1057.

[174] Squires, E. J. (1991). *Phys. Lett.* **A158**, 431.

[175] Pearle, P. & Squires, E. (1994). *Phys. Rev. Lett.*, **73**, 1.

[176] Diosi, L. (1989). *Phys. Rev.* **A40**, 1165.

[177] Ghirardi, G. C., Grassi, R. & Rimini, A. (1990). *Phys. Rev.* **A42**, 1057.

[178] Pearle, P. (1992) in *Quantum Chaos–Quantum Measurement*, eds. P. Cvitanovic, I. Percival & A. Wirzba (Kluwer, Dordrecht), p. 283.

[179] Home, D. & Majumdar, A. S. (1996). *Phys. Lett.* **A220**, 17; in *Quantum Coherence and Decoherence*, eds. K. Fujikawa & Y. A. Ono (Elsevier Science B. V.), p. 291.

[180] Kolb, E. W. & Turner, M. S. (1990). *The Early Universe*, (Addison-Wesley).

[181] Primack, J. R. (1993). *Dark Matter, Galaxies, and Large Scale Structure in the Universe* in *Particle Physics and Cosmology at the Interface*, eds. J. Pati, P. Ghose & J. Maharana (World Scientific, Singapore, 1995), p. 301.

[182] Joos, E. & Zeh, H. D. (1985). *Z. Phys.* **B59**, 223.

[183] Home, D. & Bose, S. (1996). *Phys. Lett.* **A217**, 209.

[184] Home, D. & Chattopadhyaya, R. (1996). *Phys. Rev. Lett.*, **76**, 2836.

[185] Venugopalan, A. (1994). *Phys. Rev.* **A50**, 2742.

[186] Bohr, N. (1913). *Phil. Mag.* **26**, 476.

[187] Neuhauser, W., Hohenstatt, M., Toschek, P. E. & Dehmelt, H. (1980). *Phys Rev.* **A22**, 1137.

[188] Wineland, D. J. & Itano, W. M. (1981). *Phys. Lett.* **A82**, 75.

[189] Neuhauser, W., Hohenstatt, M., Toschek, P. E. & Dehmelt, H. (1978). *Phys. Rev. Lett.* **41**, 233.

[190] Wineland, D. J., Drullinger, R. E. & Walls, F. L. (1978). *Phys. Rev. Lett.* **40**, 1639.

[191] Nagourney, W., Sandberg, J. & Dehmelt, H. (1986). *Phys. Rev. Lett.* **56**, 2797.

[192] Sauter, Th., Neuhauser, W., Blatt, R. & Toschek, P. E. (1986). *Phys. Rev. Lett.* **57**, 1696.

[193] Bergquist, J. C., Hulet, R. G., Itano, W. M. & Wineland, D. J. (1986). *Phys. Rev. Lett.* **57**, 1699.

[194] Kimble, H. J., Dagenais, M. & Mandel, L. (1977). *Phys. Rev.* **A18**, 201. Walls, D. F. (1979). *Nature*, **250**, 451.

[195] Ralls, K. S., Skocpol, W. J., Jackel, L. D., Howard, R. E., Fetter, L. A., Epworth, R. W. & Tennant, D. M. (1984). *Phys. Rev. Lett.* **52**, 228.

[196] Van Dyck, R. S. (Jr), Schwinberg, P. & Dehmelt, H. (1976). *Nature* **262**, 776.

[197] Cook, R. J. & Kimble, H. J. (1985). *Phys. Rev. Lett.* **54**, 1023.

[198] Kimble, H. J., Cook, R. J. & Wells, A. L. (1986). *Phys. Rev.* **A34**, 3190.

[199] Dehmelt, H. (1975). *Bull. Am. Phys. Soc.* **20**, 60.

[200] Home, D. & Whitaker, M. A. B. (1992). *J. of Phys.* **A25**, 2387.

[201] Einstein, A., Podolski, B. & Rosen, N. (1935). *Phys. Rev.* **47**, 777.

[202] Bohr, N. (1935). *Phys. Rev.*, **48**, 696 (Reprinted in Wheeler & Zurek [63], 1983).

[203] Schrödinger, E. (1935). *Proc. Camb. Phil. Soc.* **31**, 555–62.

[204] Schrödinger, E. (1935). *Naturwissenschaften* **23**, 807, 823, 844 (English translation in Wheeler & Zurek [63], 1983).

[205] Furry, W. H. (1936). *Phys. Rev.* **49**, 393, 476.

[206] Bohm, D. & Aharonov, Y. (1957). *Phys. Rev.* **108**, 1070.

[207] Home, D. & Selleri, F. (1991). *Rivista del Nuovo Cimento* **14**, No. 9, 1.

[208] Epstein, P. S. (1945). *Letter to A. Einstein, dated November 4*, Einstein Estate, Princeton, in Jammer, M. (1974). *The Philosophy of Quantum Mechanics* (John Wiley & Sons, New York), p. 232.

[209] Jammer, M. (1991). In *Bell's Theorem and the Foundations of Modern Physics*, eds. A. van der Merwe, F. Selleri & G. Tarozzi (World Scientific, Singapore), pp. 1–23.

[210] Bell, J. S. (1964). *Physics*, **1**, 195 (Reprinted in Wheeler & Zurek, 1983).

[211] Bell, J. S. (1966). *Rev. Mod. Phys.* **38**, 447–52.

[212] Jauch, J. M. & Piron, C. (1963). *Hev. Phys. Acta* **36**, 827.

[213] Gleason, A. M. (1957). *J. Math. & Mech.* **6**, 885.

[214] Kochen, S. & Specker, E. P. (1967). *J. Math. & Mech.* **17**, 59.

[215] Bell, J. S. (1971). In *Foundations of Quantum Mechanics* (Proceedings of the International School of Physics "Enrico Fermi", Course 49), ed. B. d'Espagnat (Academic Press, New York), pp. 171–181.

[216] Clauser, J. F., Horne, M. A., Shimony, A. & Holt, R. A. (1969). *Phys. Rev. Lett.* **23**, 880.

[217] Clauser, J. F. & Horne, M. A. (1974). *Phys. Rev.* **10**, 526.

[218] Clauser, J. F. & Shimony, A. (1978). *Rep. Prog. Phys.* **41**, 1981.

[219] Pipkin, F. M. (1978). In *Advances in Atomic and Molecular Physics*, eds. D. R. Bates & B. Bederson (Academic Press, New York).

[220] Aspect, A., Dalibard, J. & Roger, G. (1982). *Phys. Rev. Lett.* **49**, 1804.

[221] Zeilinger, A. (1986). *Phys. Lett.* **A118**, 1.

[222] Franson, J. D. (1985). *Phys. Rev.* **D31**, 2529.

[223] Shih, Y. H. & Alley, C. O. (1988). *Phys. Rev. Lett.* **61**, 2921.

[224] Ou, Z. Y. & Mandel, L. (1988). *Phys. Rev. Lett.* **61**, 50.

[225] Ghosh, R., Hong, C. K., Ou, Z. Y. & Mandel, L. (1986). *Phys. Rev.* **A34**, 3962.

[226] Horne, M. A. & Zeilinger, A. (1986). *Ann. N.Y. Acad. Sci.* **480**, 777.

[227] Bell, J. S. (1986). *Ann. N.Y. Acad. Sci.* **480**, 263.

[228] Grangier, P., Potasek, M. J. & Yurke, B. (1988). *Phys. Rev.* **A38**, 3132.

[229] Rarity, J. G. & Tapster, P. R. (1990). *Phys. Rev.* **A41**, 5139.

[230] Rarity, J. G. & Tapster, P. R. (1990). *Phys. Rev. Lett.* **64**, 2495.

[231] Franson, J. D. (1989). *Phys. Rev. Lett.* **62**, 2205.

[232] Lee, T. D. & Yang, C. N. (1961). Unpublished; discussed in Inglis, D. R. (1961) *Rev. Mod. Phys.* **33**, 1.

[233] d'Espagnat, B. (1976). *Conceptual Foundations of Quantum Mechanics* (Benjamin, London), p. 86.

[234] Six, J. (1982). *Phys. Lett.* **B114**, 200.

[235] Selleri, F. (1983). *Lett. Nuovo Cimento* **36**, 521.

[236] Datta, A. & Home, D. (1986). *Phys. Lett.* **A119**, 3.

[237] Greenberger, D. M., Horne, M. A. & Zeilinger, A. (1989). In *Bell's Theorem, Quantum Theory and Conceptions of the Universe*, ed. M. Kafatos (Kluwer Academic Press, Dordrecht), p. 73.

[238] Mermin, N. D. (1990). *Am. J. Phys.* **58**, 731; *Phys. Today* **43**, No. 8, 9; *Phys. Today* **43**, No. 12.

[239] Santos, E. (1990). *Phys. Today* **43**, No. 12, 11–13.

[240] Sawiki, M. (1990). *Phys. Today* **43**, No. 12, 11.

[241] Heywood, P. & Redhead, M. L. G. (1983). *Found. of Physics* **13**, 481.

[242] Hardy, L. (1993). *Phys. Rev. Lett.* **71**, 1665.

[243] Goldstein, S. (1994). *Phys. Rev. Lett.* **72**, 1951.

[244] Pearle, P. (1986). In *Quantum Concepts in Space and Time*, eds. R. Penrose & C. J. Isham (Oxford University Press, Oxford), p. 90.

[245] Eberhard, P. H. & Ross. R. R. (1989). *Found. Phys. Lett.* **2**, 127.

[246] Ghose, P. & Home, D. (1991). *Phys. Rev.* **A43**, 6382; In *Bell's Theorem and the Foundations of Modern Physics*, eds. A. van der Merwe, F. Selleri & G. Tarozzi (World Scientific, Singapore), p. 244.

[247] Tomonaga, S. (1946). *Prog. Theor. Phys.* **1**, 27.

[248] Schwinger, J. (1948). *Phys. Rev.* **73**, 416.

[249] Peres, A. & Ron, A. (1990). *Phys. Rev.* **A42**, 5720.

[250] Datta, A., Home, D. & Raychaudhuri, A. (1987). *Phys. Lett.* **A123**, 4; *ibid*, **A130**, 187.

[251] Squires, E. J. (1988). *Phys. Lett.* **A130**, 192.

[252] Corbett, J. V. (1988). *Phys. Lett.* **A130**, 419.

[253] MacColl, L. A. (1932). *Phys. Rev.* **40**, 621.

[254] Brouard, S., Sala, R. & Muga, J. G. (1994). *Phys. Rev.* **A49**, 4312.

[255] Büttiker, M. & Landauer, R. (1982). *Phys. Rev. Lett.* **49**, 1739. Büttiker, M. (1983). *Phys. Rev.* **B27**, 6178.

[256] Muga, J. G., Brouard, S. & Sala, R. (1992). *Phys. Lett.* **A167**, 24.

[257] Jaworski, W. & Wardlaw, D. M. (1987). *Phys. Rev.* **A37**, 2843.

[258] Hauge, E. H., Falck, J. P. & Fjeldy, T. A. (1987). *Phys. Rev.* **B36**, 4203.

[259] Hauge, E. H. & Stovneng, J. A. (1989). *Rev. Mod. Phys.* **61**, 917.

[260] Bohm, D. & Hiley, B. J. (1993). *The Undivided Universe* (Routledge, Chapman & Hall, London).

[261] Cushing, J. T. (1994). *Quantum Mechanics: Historical Contingency and the Copenhagen Hegemony* (University of Chicago Press, Chicago).

[262] Leavens, C. R. (1995). *Found. of Phys.* **25**, 229.

[263] Landauer, R. & Martin, Th. (1994). *Rev. Mod. Phys.* **66**, 217.

[264] Leavens, C. R. & Aers, G. C. (1993). In *Scanning Tunneling Microscopy III*, eds. R. Weisendanger & H.-J. Güntherodt (Springer-Verlag, Berlin), p. 105.

[265] Leavens, C. R. (1993). *Phys. Lett.* **A178**, 27.

[266] Leavens, C. R. (1994). In the Report on the *First European Workshop on Tunnelling Times*, May 1994, La Laguna, Spain (Phantoms Newsletter No. 7, October). Other references will be found in this Report.

[267] Cushing, J. T. (1995). *Found. of Phys.* **25**, 269.

[268] Ghose, P., Home, D. & Sinha Roy, M. N. (1993). *Phys. Lett.* **A183**, 267. Ghose, P. & Home, D. (1994). *Phys. Lett.* **A191**, 362.

[269] Sokolovski, D. & Baskin, L. M. (1987). *Phys. Rev.* **A36**, 4604.

[270] Sokolovski, D. & Connor, J. N. L. (1990). *Phys. Rev.* **A42**, 6512.

[271] Sokolovski, D., Bruard, S. & Connor, J. N. L. (1994). *Phys. Rev.* **A50**, 1240.

[272] Feynman, R. P. & Hibbs, A. R. (1965). *Quantum Mechanics and Path Integrals* (McGraw-Hill, New York).

[273] Iannaccone, G. & Pellegrini, B. (1994). *Phys. Rev.* **B49**, 16548.

[274] Olkhovsky, V. S. & Recami, E. (1974). *Nuovo Cimento* **22A**, 263.

[275] Lee, H. W. & Scully, M. O. (1983). *Found. of Phys.* **13**, 61.

[276] Sala, R., Brouard, S. & Muga, J. G. (1993). *J. Chem. Phys.* **99**, 2708.

[277] Garrett, C. G. B. & McCumber, D. E. (1970). *Phys. Rev.* **A1**, 305. Chu, S. & Wong, S. (1982). *Phys. Rev. Lett.* **49**, 1293.

[278] Barker, J. R. (1985). *Physica* **B134**, 22. Collins, S., Lowe, D. & Barker, J. R. (1987). *J. Phys.* **C20**, 6213. Collins, S., Lowe, D. & Barker, J. R. (1988). *J. Phys.* **C20**, 6233. Barker, J. R. (1992). In *Proc. in Physics*, Vol. 13, eds. M. Kelly & C. Weisbuch (Springer-Verlag, New York), p. 210.

[279] Baz', A. I. (1966). *Yad. Fiz.* **4**, 252 [(1967); *Sov. J. Nucl. Phys.* **4**, 229]. Rybachenko, V. F. (1967). *Yad. Fiz.* **5**, 895 [(1967); *Sov. J. Nucl. Phys.* **5**, 635].

[280] Landauer, R. (1989). *Nature* **341**, 567.

[281] Guéret, P., Marclay, E., Meier, H. (1988). *Appl. Phys. Lett.* **53**, 1617; *Solid State Commun.* **68**, 977.

[282] Esteve, D., Martinis, J. M., Urbina, C., Turlot, E., Devoret, M. H., Grabert, P. & Linkwitz, S. (1989). *Phys. Scr.* **T29**, 121.

[283] Leggett, A. J. (1984). *Phys. Rev.* **B30**, 1208.

[284] Steinberg, A. M., Kwiat, P. G. & Chiao, R. Y. (1993). *Phys. Rev. Lett.* **71**, 708. Chiao, R. Y., Kwiat, P. G. & Steinberg, A. M. (1993). *Sc. Am.* **269 (2)**, 52.

[285] Hong, C. K., Ou, Z. Y. & Mandel, L. (1987). *Phys. Rev. Lett.* **59**, 2044.

[286] Chiao, R. Y., Kwiat, P. G. & Steinberg, A. M. (1991). *Physica* **B175**, 257.

[287] Ranfagni, A., Mugnai, D., Fabeni, P., Pazzi, G. P., Naletto, G. & Sozzi, C. (1991). *Physica* **B175**, 283.

[288] Enders, A. & Nimtz, G. (1993). *Phys. Rev.* **E48**, 632.

Author index

Subject index